Laboratory EXAFS Facilities-1980
(University of Washington Workshop)

These Proceedings are dedicated to Professor Nigel J. Shevchik whose premature death prevented him from seeing the flowering of interest in laboratory EXAFS facilities that he pioneered.

AIP Conference Proceedings
Series Editor: Hugh C. Wolfe
Number 64

Laboratory EXAFS Facilities-1980
(University of Washington Workshop)

Edited by
Edward A. Stern
University of Washington

American Institute of Physics
New York   1980

Copying fees: The code at the bottom of the first page of each article in this volume gives the fee for each copy of the article made beyond the free copying permitted under the 1978 US Copyright Law. (See also the statement following "Copyright" below). This fee can be paid to the American Institute of Physics through the Copyright Clearance Center, Inc., Box 765, Schenectady, N.Y. 12301.

Copyright © 1980 American Institute of Physics

Individual readers of this volume and non-profit libraries, acting for them, are permitted to make fair use of the material in it, such as copying an article for use in teaching or research. Permission is granted to quote from this volume in scientific work with the customary acknowledgment of the source. To reprint a figure, table or other excerpt requires the consent of one of the original authors and notification to AIP. Republication or systematic or multiple reproduction of any material in this volume is permitted only under license from AIP. Address inquiries to Series Editor, AIP Conference Proceedings, AIP.

L.C. Catalog Card No. 80-70579
ISBN 0-88318-163-0
DOE CONF- 800487

FOREWORD

This Proceedings is the outcome of a "Workshop on Laboratory EXAFS Facilities and Their Relation to Synchrotron Radiation Sources" held in Seattle, Washington, April 28-30, 1980. The Workshop was held to evaluate the capabilities of Extended X-ray Absorption Fine Structure (EXAFS) facilities using conventional bremstrahlung sources, which are suited for use in one's own laboratory, and to assess their relationaship to the EXAFS facilities at national synchrotron radiation sources.

In the last decade, after it was shown that EXAFS can be utilized to obtain information on the atomic arrangement of materials*, EXAFS measurements have experienced a phenomenal growth. The feature of EXAFS that makes it attractive is its capability to measure the atomic arrangement around a chosen atom type independent of whether the material is crystalline or not. This new technique has made feasible structure determination on systems that were not amenable to the more standard techniques.

The demand to do EXAFS measurements gave an impetus to the development of synchrotron radiation sources. At the sources, EXAFS facilities were constructed and instrumentation developed which greatly facilitated the measurements. The EXAFS facilities at the Stanford Synchrotron Radiation Laboratory, the first of such facilities, made accessible EXAFS measurements to the general scientific community leading to a spectacular growth in the application of the technique. As the scientific community became acquainted in this manner to the usefulness of EXAFS the demand quickly outstripped the available facilities. For this and other reasons described in the Proceedings there began a trend to develop techniques for doing EXAFS measurements in the laboratory reviving a technology that had almost disappeared. The 1930's saw the development of the basic technology being used today. However, the modern instruments described in the Proceedings take full advantage of computer technology and modern electronics which give them capabilities that would make the originators of the technology envious.

It seemed appropriate to convene a workshop to make a studied effort to evaluate the relationship between laboratory and synchrotron radiation EXAFS facilities and to assess what may be their relative strengths and weaknesses. These questions are ones that may face many research groups and funding officials in industry, universities, and government, as the assessment is made of how to invest both time and funds in the growing field of EXAFS. Both technical and policy questions are involved. The two questions are dependent on one another and it seemed appropriate to discuss both at the same meeting.

*D.E. Sayers, E.A. Stern, F. Lytle, Phys. Rev. Letters $\underline{27}$, 1204 (1971).

To that purpose many of the people active in the field of EXAFS, were invited to participate. The list of the attendees is given in the Appendix. As can be noted, the spectrum covered both those in research and those influential in policy making. A summary of the conclusions from the Workshop is presented in the next section.

The Workshop was organized into a series of plenary presentations where the present state of the various elements of a laboratory EXAFS facility were given. Chapters 1 to 7 cover these presentations. Then four separate workshops were convened to cover the topics of Sources, Crystals and Focusing, Detectors, and Hardware and Software, chaired by B. R. Stults, D. W. Berreman, D. Sandstrom and P. Georgopoulos, respectively. The presentations of these workshops summarized by their respective chairmen and contributions by panel members and others are given in Chapters 8 to 11. The final presentation of the Workshop considered the "Relation of Roles of EXAFS Facilities in Laboratory and at Synchrotron Radiation Sources" and also constitutes the last chapter of the Proceedings, Chapter 12.

Inherent in the nature of a Proceedings is the problem of the continuity and coherence of the various contributions. The author giving his presentation has to make some assumptions of the topics that will or will not be covered by the other authors. This inevitably leads to repetition in some areas and gaps in others. To help overcome this problem I have taken the liberty of adding a more generous number of editor's notes than is usually the case for a Proceedings. To distinguish the editor's notes from those of the authors', they are referred to by lower case superscripts starting from the end of the alphabet and going backwards. I hope this ordering does not reflect on the notes' contents but it assures that the reference is unique and will not be confused with that of any of the authors. The contributions are divided into chapters and the editor's notes referenced in each chapter are placed at the end of that chapter.

The convening of the Workshop would not have been possible without the generous financial help of Battelle Seminars and Conferences, Seattle, Washington, and the Monsanto Company. Dr. B. R. Stults is particularly deserving of thanks for arranging the support from the Monsanto Company. The initial encouragement and support of Provost George Beckmann and Dean Ronald Geballe of the University of Washington were the essential catalysts that led to the organizing activities which are culminating in these Proceedings.

It would have not been possible to have organized the Workshop and these Proceedings without the help of many people. First, the members of the program committee gave very wise and important advice besides contributing substantially to the Workshop by their participation. They are: John Baldeschwieler, Arthur Bienenstock, Gabrielle Cohen, Gordon Knapp, and B. R. Stults. I also want to give thanks to Donald Sandstrom for his particular helpfulness and to other

chairmen of the individual workshops Dwight Berreman and
P. Georgopoulos. My graduate students, associates and post docs
were leaned on particularly heavily because of their close proximity.
I am appreciative not only of their important help and participation
but also by their willingness to be helpful. They are: Charles
Bouldin, Grant Bunker, Edward Keller, Kyung-Ha Kim, Kun-quan Lu,
Yaacov Azoulay, Bruce Bunker, and W. T. Elam. Mitzie Johnson
deserves a special thanks. She handled the administrative details
and typed the corrections to the Proceedings helping to organize
them into their final form. She managed to get married during the
same time, though the Workshop takes no credit or responsibility
for that. Finally, I am pleased to express my appreciation to the
participants of the Workshop, a list of which is presented in the
Appendix, whose contributions, lively interest and discussions constituted the essential element in making the Workshop the success
that it was.

            Edward A. Stern
            Department of Physics
            University of Washington

TABLE OF CONTENTS

Summary		1
Chap. 1.	GENERAL CONSIDERATIONS FOR A LABORATORY EXAFS FACILITY G.S. Knapp and P. Georgopoulos	2
Chap. 2.	AN X-RAY SOURCE FOR A LABORATORY EXAFS FACILITY G.R. Fisher	21
Chap. 3.	EVALUATION OF FOCUSING MONOCHROMATORS FOR AN EXAFS SPECTROMETER S.M. Heald	31
Chap. 4.	DETECTORS FOR LABORATORY EXAFS FACILITIES E.A. Stern	39
Chap. 5.	LABORATORY EXAFS FACILITY HARDWARE AND SOFTWARE W.T. Elam	51
Chap. 6.	INSTRUMENTAL ASPECTS OF EXELFS ANALYSIS IN THE ELECTRON MICROSCOPE D.E. Johnson, S. Csillag and E.A. Stern	63
Chap. 7.	COMPARISON OF LABORATORY AND SYNCHROTRON RADIATION EXAFS FACILITIES R. Haensel	73
Chap. 8.	WORKSHOP ON X-RAY SOURCES	
	SUMMARY B. Ray Stults	84
	ROTATING ANODE X-RAY SOURCE John Holbin	88
	X-RAY SOURCE FOR EXAFS D.G. Hempstead	89
	USE OF A SCANNING ELECTRON MICROSCOPE AS AN X-RAY SOURCE FOR EXAFS H.W. Deckman, J.H. Dunsmuir, G. Via	91
	SOFT X-RAY SOURCES J. Azoulay	93

LASER-EXAFS: LABORATORY EXAFS WITH A NANOSECOND PULSE
OF LASER-PRODUCED X-RAYS
P.J. Mallozzi, R.E. Schwerzel, H.M. Epstein.............. 96

Chap. 9. WORKSHOP ON CRYSTALS AND FOCUSING

WORKSHOP ON CRYSTALS AND FOCUSING
D.W. Berreman.. 100

JOHANN AND JOHANSSON FOCUSING ARRANGEMENTS: ANALYTICAL
ANALYSIS
Kun-quan Lu and E.A. Stern............................... 104

X-RAY ABSORPTION SPECTROMETER WITH A DISPERSIVE AS WELL
AS FOCUSING OPTICAL SYSTEM
Tadashi Matsushita....................................... 109

ADJUSTABLE CRYSTAL BENDING APPARATUS
S.E. Crane... 111

APPLICATION OF MIRRORS FOR FOCUSING X-RAYS
R.C. Gamble.. 113

CRYSTALS AND FOCUSING GEOMETRY
R.S. Emrich and J.R. Katzer.............................. 117

Chap. 10. WORKSHOP ON DETECTORS

SUMMARY
D.R. Sandstrom... 122

EXAFS SPECTROSCOPY USING A FLAT CRYSTAL AND LINEAR
DETECTOR
R.C. Gamble.. 123

POLARIZATION CONSIDERATIONS IN THE FLUORESCENCE
DETECTION OF EXAFS
D.R. Sandstrom and J.M. Fine............................. 127

OPTIMIZATION OF THE X-RAY DETECTION SYSTEM FOR EXAFS
Y. Yacoby.. 129

DETECTION SYSTEMS
R.J. Emrich and J.R. Katzer.............................. 131

Chap. 11. WORKSHOP ON AUTOMATING AN EXAFS FACILITY:
HARDWARE AND SOFTWARE CONSIDERATIONS
P. Georgopoulos, D.E. Sayers, B. Bunker, T. Elam,
W.A. Grote... 134

Chap. 12. WORKSHOP ON ROLES OF IN-LABORATORY AND SYNCHROTRON
RADIATION EXAFS FACILITIES

RELATION OF ROLES OF IN-LABORATORY AND SYNCHROTRON
RADIATION EXAFS FACILITIES
Arthur Bienenstock.. 150

APPLICATION OF EXAFS IN CHEMISTRY
John D. Baldeschwieler.. 155

RELATION OF ROLES OF IN-LABORATORY AND SYNCHROTRON
RADIATION EXAFS FACILITIES
B. Ray Stults... 158

ROLES OF LABORATORY VERSUS SYNCHROTRON RADIATION EXAFS
FACILITIES
Edward A. Stern... 160

Appendix: List of Participants.............................. 163

SUMMARY

The Workshop showed that it is possible to build a laboratory EXAFS facility using a fixed anode bremstrahlung x-ray generator with a photon flux of around $1-3 \times 10^6$ photons/sec in the resolution bandwidth of about 5 eV. Such a facility at the University of Washington using Si(400) monochromating crystals was demonstrated at the Workshop. With the innovations suggested at the conference, in particular, the development of improved detectors as suggested by Y. Yacoby, and crystals better matched to the resolution width, the intensity could be increased by perhaps a factor of 10. With the use of rotating anode sources in place of the fixed anode another factor of 15 can be attained. Thus it appears possible to build laboratory EXAFS facilities with effective intensities above 10^8 photon/sec in a resolution width of 5 eV. These intensities are all quite suitable for measuring EXAFS spectra in concentrated samples, and the highest intensities which the future generation of laboratory EXAFS facilities may attain are competitive with that at the present generation of synchrotron sources. Although the future generation of synchrotron radiation sources will be several orders of magnitude more intense, it was generally agreed that the intensities presently available in laboratory EXAFS facilities make them completely competitive for measurements on concentrated samples and moderately dilute ones. The synchrotron sources have decided advantages in measuring very dilute samples, higher energy resolution, polarization effects, and the high energy spectrum above 20 KeV. At the moment the soft x-ray region of EXAFS measurements is exclusively being covered by synchrotron sources but the ongoing improvement of electron energy loss instrumentation in electron microscopes described in Chapter 6 may open up this region to laboratory EXAFS study.

The possibility of building an EXAFS facility under $100,000 using fixed anode sources makes quite practical the possibility of a large growth in the number of such laboratory facilities. This may have the consequence of concentrating the routine EXAFS measurements in the laboratory, freeing synchrotron sources for the more difficult ones. The growth of the number of laboratory EXAFS facilities should increase the quality of the research in the field by stimulating better training of researchers and innovativeness in the field, both in the laboratory and at the synchrotron radiation sources.

E.A. Stern

GENERAL CONSIDERATIONS FOR A LABORATORY EXAFS FACILITY*

G. S. Knapp and P. Georgopoulos
Materials Science Division
Argonne National Laboratory
Argonne, IL 60439

ABSTRACT

In this paper a complete laboratory system capable of making high quality EXAFS measurements will be described. A discussion will be given of the design of an EXAFS spectrometer, how to achieve an optimum combination of intensity and resolution, the range of energies which are practical to take EXAFS patterns, the elimination of errors caused by characteristic lines, detector systems, and the computer requirements to obtain and analyze the data.

INTRODUCTION

In recent years the Extended X-ray Absorption Fine Structure (EXAFS) technique has become an important tool for structural analysis.[1-4] The technique yields information about the local atomic environment of the particular element being studied. An EXAFS experiment requires a very precise measurement of the x-ray absorption coefficient as a function of energy. Figure 1 shows typical EXAFS patterns from a number of iron containing compounds. The EXAFS information is contained in the complex oscillatory behavior from 50–1000 eV above the absorption edge of the element. These oscillations arise from scattering and self-interference of the photoemitted electrons by the surrounding atoms.

In order to analyze a pattern, such as shown in Fig. 1, one writes

$$\mu(E) = \mu_o(E) [1 + \chi(E)] \qquad (1)$$

where $\mu_o(E)$ is a smooth properly normalized background. If one uses the relation between the momentum and the energy of the outgoing electron, $\hbar^2 k^2/2m = E-E_o$, where E_o is close to the edge energy, then

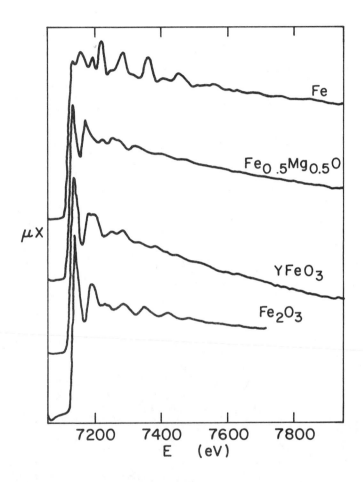

Fig. 1. The x-ray absorption coefficient (multiplied by the thickness) as a function of energy in the vicinity of the Fe K absorption edge for a number of iron-containing compounds.

$$\chi(k) = \frac{m}{4\pi h^2 k} \sum_j \frac{N_j}{R_j^2} t_j(k) \exp(-2R_j/\lambda)$$
$$\times \sin[2kR_j + \phi_j(k)] \exp(-2k^2\sigma_j^2) , \qquad (2)$$

where N_j is the number of atoms in the jth coordination sphere, R_j is the average radius of this sphere, $t_j(k)$ is the magnitude of the backscattering amplitude for scattering off the atoms in the jth sphere and λ is the mean free path for inelastic scattering of the photoelectron. The exponential factor containing σ_j^2 is a Debye-Waller-type term and $\phi_j(k)$ is a phase shift. From theory or experiment many of the parameters of this equation (such as $t_j(k)$, $\phi_j(k)$ and λ) can be determined so that N_j, R_j and σ_j can be considered the information the EXAFS technique allows us to determine.

In a dilute sample, the information in $\chi(k)$ is a very small fraction of the total absorption, hence high count rates are necessary to achieve the necessary statistical accuracy. In addition, great care must be taken to avoid even small systematic errors. Until recently, because of this need for high count rate it was thought that the only practical way to perform an EXAFS experiment was to utilize a synchrotron x-ray source. However, more and more groups have become interested in a facility to make such measurements in the laboratory. This conference is a result of that interest.

In this paper we will describe a complete laboratory system capable of making high quality EXAFS measurements in relatively short times. We will mainly discuss the system which we have developed. However, mention should be made of the pioneering work of the Stonybrook Group, lead initially by the late Nigel Shevchik and later by Gabrielle Cohen. Many of their ideas have been incorporated in our design.

Our system, which has been undergoing continuous development for approximately 4 years, is now capable of producing count rates between 10^6 to 10^7 photons per sec in the 5-15 keV range with better than 15 eV resolution. We have developed a method where errors due to characteristic lines can be almost completely eliminated. Our system is computer controlled, using a LSI-11 based system (see Georgopoulos et al. in this Volume).[2] This computer system runs both the experiment and allows complete analysis of the data. Our system is capable of performing much of the experiments currently being carried out at a synchrotron.

X-RAY SOURCES

The choice of an x-ray source for an EXAFS experiment is largely governed by one's budget. If one has available a conventional x-ray generator, it is possibe to perform many experiments without the expense and trouble of a rotating anode source. Clearly, the rotating anode, with approximately a factor of ten increase in flux, is preferable if one can afford the roughly $100,000 price. We utilize an Elliott (Model GX-21) rotating anode generator, equipped with a gold anode and operating at a maximum of 300 mA of anode current at voltages up to 50 kV. By reducing the distance between the filament and the anode, voltages as low as 10 kV can be achieved at full current. The low voltage operation is necessary to completely eliminate high energy x-rays passed by the monochromator. Using a conventional x-ray source, fluxes close to 10^6 c/sec are possible and with a rotating anode 10^7 c/sec can be achieved.

1. The spectrometer

In one of our earlier papers,[5] we describe how a conventional diffractometer could be adapted to make transmission EXAFS experiments. However, this method is mechanically complex and it is not applicable to fluorescence detection.

We have developed[6] a fully focusing spectrometer which is shown in Fig. 2. Cohen et al.[7] have described a similar design. This apparatus is designed to employ either Johann or Johansson cut monochromator crystals, with a Rowland circle radius of 20 cm. The scattering angle can be varied between 30-70°; the range where an optimum combination of resolution and intensity can be realized.

Both the crystal, as well as the receiving slit, move on the Rowland circle at the same time in order to satisfy the diffraction condition. Figure 2 illustrates the principles of operation of the spectrometer. The point A is placed directly under the anode image. The point C, above which the focusing crystal is located, is moveable by means of a lead screw and a computer controlled stepping motor. The arms RA, RC and RS are all equal to r, the Rowland circle radius of the monochromator crystal. The arms BA and BS are equal and their function is to keep the angles $\angle ACR = \angle RCS = \pi/2 - \theta$.

This spectrometer has a number of advantages for an EXAFS experiment. First, the entire set of lever arms comprising the spectrometer are assembled on a rigid flat 32 mm thick anodized aluminum plate. A robust specimen stage, capable of carrying heavy loads (such as cryostats, furnaces, detectors, etc.) is carried behind the receiving slit and made to slide on the 32 mm plate with little friction. This is a big advantage over a conventional

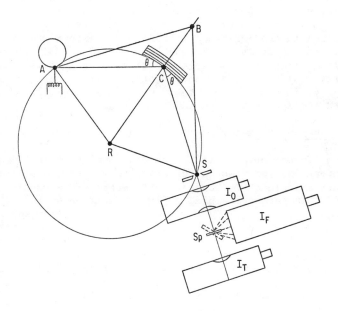

Fig. 2. EXAFS spectrometer schematic, illustrating the geometrical principle, as well as the sample-detector arrangement: A-anode, C-crystal, S-receiving slit, I_o-beam monitor detector, I_F-fluorescence detector, I_T-transmission detector, Sp-specimen.

diffractometer, which has a rather limited load capacity on its receiving arm.

Second, because of the use of a receiving slit, the energy resolution of the spectrometer is independent of the crystalline perfection of the monochromator crystal and depends only on its absorption coefficient, the geometrical perfection of its ground surface and the widths of the anode image and receiving slit, all of which are more easily controllable. If the crystal is perfectly ground and opaque to x-rays and vertical defocusing is ignored, then the FWHM range $\Delta\theta$ about the mean angle θ, through which the crystal can reflect, is given by Knapp et al.[5]

$$\Delta\theta = \frac{W_a + W_s}{4r \sin\theta} \quad . \qquad (3)$$

Here, W_a and W_s are the projected anode image and receiving slit width, respectively.[y] If we now consider the monochromator to be highly mosaic and a crystallite has its diffracting planes at an angle to the surface, diffraction will of course take place, but the diffracted beam will not pass through the receiving slit unless the crystallite is tilted by no more than $1/2\Delta\theta$. Obviously, if the mosaic spread is either much larger or much smaller than $\Delta\theta$, the brightness will be reduced drastically. This last point has been discussed by Lytle et al.[3] and, even though they consider flat crystal monochromators, their arguments apply to curved crystals as well.

2. Monochromators[x]

The key element in any EXAFS spectrometer is the curved crystal monochromator. As of this date, we have not found a commercial source of satisfactory bent crystals. Instead, following Stern and Lu*, we have found it is much more profitable to produce our own. Using a bending jig, which allows us to adjust the bending radius of the crystal when it is mounted on the spectrometer, we have achieved both good brightness and resolution with a Si(400) crystal in the Johann configuration. We are looking for a good source of Ge crystals, which will be brighter and have better resolution.

The various factors affecting both the resolution and brightness of the crystals will be examined below.[w] From the equation derived in Knapp et al.[5] and from simple geometrical calculations of the effects of the depth of penetration of the x-rays into the crystal and the effects of vertical divergence of the beam, one can obtain

*This information was transmitted at the Workshop.

$$\Delta\theta = [8r\sin\theta]^{-1} \left[(W_a + W_s)^2 + \left(\frac{2\ell n2 \cos\theta \sin\theta}{\mu}\right)^2 + \left(\frac{h^2}{8r\cos\theta}\right)^2\right]^{1/2} \quad , \quad (4)$$

assuming that the broadening factors add in quadrature. Here, W_a is the projected width of the anode image, W_s the width of the receiving slit, μ is the absorption coefficient of the crystal, h is the height of the anode image and the receiving slit (which we take to be equal), r is the radius of the Rowland circle and θ is the angle of diffraction. This equation is only approximate[w] since the various broadening factors are not Gaussian.[8] It is, however, quite accurate for the purpose of evaluating monochromator crystals on a relative basis and obtaining a reasonable estimate for the expected resolution.

From Eqn.(4) we can calculate the energy resolution. However, this equation is relatively complex and by making a number of substitutions and approximations, we can find a simpler form. Over the usual range of energies we have

$$\cos\theta \approx 1$$

$$E(eV) = \frac{C}{2d\sin\theta} \quad , \quad C = 12,396 \text{ eV·Å}$$

$$\mu \approx \frac{K}{E^3}$$

and substituting in Eq. (4) we can obtain

$$\Delta E \approx \frac{E^3(2d)^2}{8rC^2} \left[(W_a + W_s)^2 + \left(\frac{C\ell n2 E^2}{Kd}\right)^2 + \left(\frac{h^2}{8r}\right)^2\right]^{1/2} \quad . \quad (5)$$

This equation shows that ΔE depends on $(2d)^2$, so by using high index plane crystals adequate resolution can always be achieved. In addition, the 2nd term in the brackets has a very strong energy dependence so a crystal with a large enough K must be chosen for high energy work. Ge has a high absorption coefficient so this term is negligible, but for Si and LiF this is not the case. The optimum attainable resolution with h = 1 cm, $W_a + W_s$ =0.01 cm and v=2.0cm is shown in Fig. 3 for a selection of monochromator crystals as a function of energy. The actual resolution obtained with a particular spectrometer and monochromator crystal may be worse. This is partly due to the fact that, as previously mentioned, the broadening factors in Eq. (4) are assumed Gaussian in shape, which is not strictly true. More importantly, any deviations from the ideal spectrometer geometry assumed in Eq. (4) can be a source of

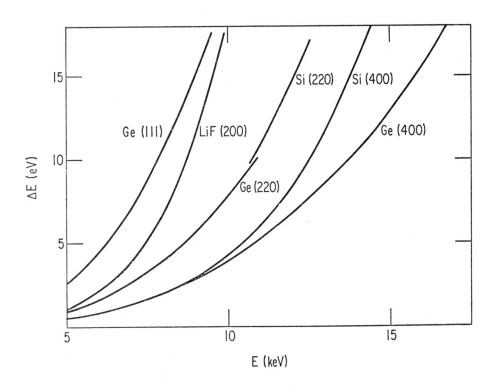

Fig. 3. Energy resolution vs energy for various monochromator crystals. Si(111) and Si(220) below 10.5 keV (not shown for clarity) are approximately the same as Ge(111) and Ge(220), respectively.

serious deterioration in energy resolution. Such deviations may be due to lack of parallelism in the x-ray focal spot-crystal-receiving slit system, imperfectly ground and/or bent monochromators, etc. To illustrate the magnitude of these effects, let us assume that during preparation, the surface normal and the normal of the diffracting planes of the monochromator crystal differed by $0.2°$ in the horizontal plane. When the crystal is aligned for maximum diffracted intensity, its face will deviate from the focusing circle by up to ± 17.5 μm at its edges and the broadening due to this variation will be comparable to that due to the size of the x-ray focal spot and receiving slit ($W_a = W_s = 50$ μm was used for the calculation plotted in Fig. 3). It is obvious then that in order to achieve resolutions such as those shown in Fig. 3, the spectrometer must be accurately constructed and aligned and the monochromator crystals must likewise be carefully oriented, cut and ground. The best resolution we have been able to achieve in our laboratory, using a Si(400) crystal, was approximately 40% worse than that predicted by Eq. (4). Since approximately 20 eV resolution is the maximum allowable for a successful EXAFS experiment, Fig. 3 and Eq. (4) show that it will be difficult to perform in-laboratory EXAFS experiments at energies above 20 keV. One must utilize Ge crystals, with very small 2d spacings, which of course mean that the count rates will be low.

Finally, we will consider briefly certain factors affecting the brightness of various crystals. It can be shown[9] that for ideally imperfect crystals (all crystals when bent to a radius of 20 cm can be considered ideally imperfect) the ratio of intensities diffracted by any two crystals at any two energies E_1 and E_2 is given by

$$\frac{I_1(E_1)}{I_2(E_2)} \approx \frac{E_2^3 \mu_2 |F_1|^2 v_2^2}{E_1^3 \mu_1 |F_2|^2 v_1^2} \qquad (6)$$

where $\mu_1(\mu_2)$, $F_1(F_2)$ and $v_2(v_2)$ are the absorption coefficient, structure factor and unit cell volume, respectively, and Lorentz-polarization corrections have been ignored, as they vary only slowly over the typical range of angles. Table 1 shows the relative brightness of various crystals [compared with LiF(200)]. Note, that in the absence of absorption edges, the product $E^3 \mu$ is approximately constant, hence, the relative brightness shown in Table 1 is independent of energy except in the case of Ge. It must be emphasized that Eq. (6) can only be used as a rough guide, since such uncontrollable factors as the mosaic spread of the crystal are not included.

Table 1. Brightness of various monochromator crystals relative to LiF(200).

Crystal	d-spacing	Relative brightness
LiF(200)	2.0135	1
Ge(111)	3.2660	0.28
Ge(220)	2.0000	0.42
Ge(400)	1.4142	0.31 (0.04)*
Si(111)	3.1353	0.16
Si(220)	1.9200	0.21
Si(400)	1.3576	0.15

*Above the Ge K absorption edge (11103 eV).

3. Detector systems[V]

There are several types of x-ray detectors operating on a variety of principles. Here, we will only examine the ones that seem to be suitable for an EXAFS spectrometer. Generally, a detector is needed to monitor the incoming beam, (I_o detector), and another detector to measure either the transmitted or fluorescence x-rays.

Let us first consider the I_o detector. The x-ray beam striking the detector will have an intensity between 10^5 to 10^7 photon/sec. The detector should absorb between 5 to 20% of these photons, depending on the experiment, therefore, we would like a detector system with low noise and reasonably linear up to 10^6 photon/sec.

Since this detector must only absorb a portion of the beam, clearly a gas filled detector, either an ionization or a proportional detector, is suitable. We have not utilized ionization detectors in our apparatus, but our experience has shown that a suitably designed proportional detector works rather well. In Fig. 4 we show a block diagram of our detector system. This system allows variable gas pressure (either argon or neon mixed with a suitable quench gas) to vary the amount of absorption and is capable of detecting x-rays at rates up to 10^6 c/sec. However, it is only linear up to about 4×10^5 c/sec. With enough

12

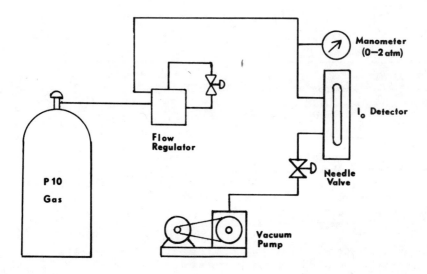

Fig. 4. Variable pressure I_o detector system.

care to reduce dark current fluctuations, ionization detectors would probably be superior.

The final detector in a transmission experiment has approximately the same count rate requirements as the I_o detector, except it must absorb the total beam. A xenon filled proportional detector works fairly well in this application.

For fluorescence detection in the laboratory, a NaI detector seems to be the best choice, since high count rates are not encountered. However, NaI detectors have very poor pulse height resolution and a filter system of the type described by Stern et al.[10] is often necessary. A solid state detector would be superior, provided it can be made fast without degrading its energy resolution inordinately. Another possibility is the gas scintillation proportional detectors discussed by Polycarpo et al.[11].

AUTOMATION[u]

The necessity of some form of computational capability for data reduction and analysis hardly deserves special emphasis. What we are primarily interested in here are the advantages that a micro- or mini-computer installation has over large-scale computers. Certain advantages are obvious: A small micro- or mini-computer is inexpensive enough to be used in a fully dedicated mode and it is relatively easy to program. In addition, it can be easily interfaced to the EXAFS system. Of course, some form of automation of the experiment is necessary in any case, and certain laboratories use daisy-chained scaler-stepping motor-printing devices or multi-channel analyzer systems for data acquisition. However, the use of computer control makes the system so much more versatile, as many parameters can be varied, if appropriate, during the course of a measurement. Additionally, digital control of the generator current for x-ray flux stabilization is far more precise, stable and safe for the equipment than the analog feed back control we have utilized in our laboratory in the past.

One other advantage that small dedicated computers have over large central installations is flexibility and large throughput. This last item may seem absurd at first, until we examine the computational needs of EXAFS data reduction and analysis. The present stage of theoretical development of EXAFS parameters (backscattering amplitudes, phase shifts, edge shifts, background, etc.) coupled with the intrinsically low informational content of an EXAFS pattern makes the analysis of raw data empirical in certain stages. At the same time, the computational power required is rather modest, as simple numerical manipulations are involved.

SIGNAL-TO-NOISE CONSIDERATIONS[t]

In transmission, the EXAFS oscillations are ~10% of the absorption coefficient due to the K-shell absorption/μ_K. From $\mu x = \ln I_0/I$, it is easy to show that $\delta\mu(E)$, the statistical fluctuation in $\mu(E)$, is related to the number of counts N_0 incident on the sample and the total absorption coefficient by

$$\frac{\delta\mu(E)}{\mu_K(E)} \sim N_0^{-1/2} \frac{\mu(E)}{\mu_K(E)}, \qquad (7)$$

when the sample is approximately of optimum thickness $\mu x \approx 2$ just above the edge. The left-hand side of Eq. (7) must be 0.01 or less (signal noise, $S/N \geq 100$) in order to accurately determine the near-neighbor distance, number of atoms, etc. We can, therefore, use Eq. (7) to calculate the absorber concentration for which $\delta\mu/\mu_K \leq 10^{-2}$ is obtained. These minimum concentrations are shown in Table 2 for the various experimental methods. The synchrotron facility is clearly superior in terms of flux, but the user has a severely limited duty factor on the facility.

If the fluorescent technique is used, much more dilute experiments can be carried out. Jaklevic et al.[12] were the first to describe the method using synchrotron radiation. Del Cueto and Shevchik[13] described how this method can be used in the laboratory. It can be shown that the fluorescence intensity is given by

$$I(E_F) = I_0(E)\varepsilon \frac{\Omega}{4\pi} \frac{\mu_K(E)}{\mu(E) + \mu(E_F)\sin\beta}$$
$$\times \left(1 - \exp\left\{-\left[\frac{\mu(E)}{\sin\beta} + \mu(E_F)\right]\right\} d\right). \qquad (8)$$

Here, $I(E_F)$ is the detected characteristic fluorescence intensity, $I_0(E)$ the intensity of the incident x-ray, d the thickness of the sample, $\sin\beta$ the angle between the plane of the sample and the incident beam, ε the fluorescence yield, and $\Omega/4\pi$ the solid angle subtended by the detector. The fluorescence yield ε varies from 0.02 at $Z = 10$(Ne) to 0.75 at $Z = 42$(Mo). Since it is relatively independent of energy, it is clear from Eq. (8) that this method will allow detection of the EXAFS oscillations. Using Eq. (8), when $\sin\beta$ is small and $d \gg [\mu(E)/\sin\beta + \mu(E_F)]^{-1}$, we can show that for fluorescence detection

Table 2. Minimum concentrations for various experimental methods using the transmission techniques.

Method	Count Rates (sec^{-1})	μ_K/μ for a 20 hr run	Examples
Flat crystal	~10^3	0.7	elements
Curved crystal with 1 kW generator	~10^6	~0.02	~.4% Fe in H_2O
Curved crystal with rotating anode	~10^7	~0.01	~0.2% Fe in H_2O
SSRL	10^8-10^{10}/time on machine (%)		

$$\frac{\delta\mu(E)}{\mu_K(E)} \sim N_o^{-1/2} \frac{\varepsilon\Omega}{4\pi}^{-1/2} \frac{\mu(E)}{\mu_K(E)}^{1/2} . \qquad (9)$$

Comparing Eqs. (7) and (9), it is apparent that for fluorescent detection the noise-to-signal ratio varies with the square root of μ/μ_K and not linearly as in absorption detection.

Unfortunately, unless special techniques are used, Eq. (9) is much too optimistic about the advantages of the fluorescence technique. In a dilute sample a background signal, due to both elastic and Compton scattering, will greatly decrease the sensitivity of the technique. Nevertheless, by using the filter described by Stern and Heald,[10] the fluorescence method is far superior to the transmission technique when $\mu_K/\mu < 0.05$.

LIMITATIONS AND DIFFICULTIES

This is the most important section of this report. While most of the discussion of laboratory EXAFS seems to concentrate on brightness and resolution, in our experience it is systematic errors and/or the limited information content that cause most experiments to be unsuccessful. In Table 3 we list problems in their order of importance, which must be dealt with before a successful EXAFS experiment can be completed.

The first, the limited informational content, stems from the restricted k range of the experiment, limiting the number of parameters that can be uniquely determined. Fortunately, the problem can be anticipated before the experiment is even carried out. What one must do is make computer simulations of EXAFS spectra (using either experimentally or theoretically determined phase shifts and amplitudes), of the possible atomic environments of the system being studied. Generally, one attempts several models which are reasonably possible, and if the spectra of the various models are different enough, one then attempts the experiment. In some cases, we have found the models which are quite different chemically can give remarkably similar EXAFS patterns, hence the experiment will not uniquely determine which model is correct.

The second problem usually comes about from a combination of nonlinearity of the detector systems and characteristic lines in the x-ray spectra. Fortunately, there is a way to largely eliminate the problem of characteristic lines. First, one must use a monochromatic beam, i.e., operate the x-ray generator at low enough voltage to eliminate higher order reflections and, secondly use some type of feedback system to stabilize the flux detected by the I_o detector. It is very convenient to use the computer to supply the appropriate signal to control this flux. An example of the "glitches" that can occur in the presence of characteristic

Table 3. Limitations and difficulties of the EXAFS technique.

1. Limited informational content.
2. Glitches.
3. Sample preparation.
4. Resolution.
5. Statistics.

lines is given in Fig. 5 , which shows the EXAFS spectrum of ZnO, accumulated at constant incident flux (upper curve) and constant generator power (lower curve). Admittedly, this is a rather pathological case, and one should avoid such extremes by changing the target element. At the 0.1% level of accuracy, even minor impurity lines, which are always present, can be a serious problem, however, at least with proportional counters which run close to their limits. Ionization detectors would probably be better in this respect but, at count rates of a few hundred thousand cps, they also run close to their lower limit where nonlinearities due to dark current are expected.

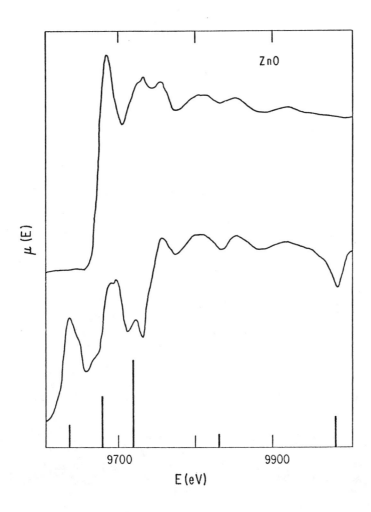

Fig. 5. EXAFS spectrum of ZnO measured with (upper curve) and without (lower curve) flux stabilization. Vertical bars indicate the positions of major spectral lines.

The third problem, that of sample preparation, can be quite difficult. For a transmission sample, the samples must be thin, and very homogeneous, or major amplitude distortions will occur.

The fourth and fifth problems are related. Resolutions of 20 eV or better are required, otherwise the first few EXAFS oscillations will be washed out. It is always possible to achieve the necessary resolution with a crystal with a small enough 2d spacing or by using a flat or double crystal monochromator; however, the flux may be inadequate (fluxes below 5×10^5 c/sec are impractical and one should go to a synchrotron).

SUMMARY

This paper was not intended to give a detailed discussion of all factors governing the in-laboratory EXAFS technique, but rather an overview of the possibilities and the promises. Our system, which has been in a continuing state of development for 4 years, is now capable of performing quite difficult experiments. For example publishable results have been obtained on a sample of Nb_3Ge only 1µ thick[14], dilute uranium solutions[15] (we studied the EXAFS above the LIII edge) and samples as dilute as 0.1% Cu in Al. Fluxes of 4×10^6 c/sec with a Si(400) Johann crystal have been achieved in the 7-10 keV region. Using a properly bent Ge(220) crystal, fluxes of at least 10^7 c/sec will be realized in this energy range. If a good source of Johansson ground crystals could be found, another 3-fold increase may be expected. Fluxes of this magnitude will demand ionization detectors of the type discussed by Stern in this conference and we are currently developing such detectors. With the development of proper crystals, spectrometer, detectors and the like, in-laboratory EXAFS, is now practical using only a conventional x-ray generator, since fluxes of 10^6 c/sec allow many experiments to be carried out in an hour or less.

The difficult experiments which demand fluxes of 10^6 c/sec or greater, also will demand great care to eliminate systematic errors. Even with high fluxes, reducing the magnitude of the systematic errors to below 0.1% is difficult. The use of fast detectors and flux stabilization techniques is very important. Equally important are the care in sample preparation and, of course, in analysis of the data.

Clearly the EXAFS technique is an important new tool for the materials scientist, chemist or biologist. The advent of the in-laboratory model of EXAFS experiments opens new doors to a wider range of scientists.

Acknowledgements

Many people were instrumental in the development of our techniques. At Argonne, F. Y. Fradin, H. Chen (now at the University of Illinois), and T. E. Klippert have played important roles in the development of the facility. The work of the Stonybrook group has also been important, particularly in introducing us to the possibility of new fluorescence techniques in the laboratory and also to the design of the focusing spectrometer.

References

1. G. S. Knapp and F. Y. Fradin in Electron and Positron Spectroscopies in Materials Science and Engineering, eds. O. Buck, J. K. Tien and H. L. Marcus (Academic Press, New York, 1979) pp. 243-274.
2. E. A. Stern, Phys. Rev. B $\underline{10}$, 3027 (1974).
3. F. W. Lytle, D. E. Sayers and E. A. Stern, Phys. Rev. B $\underline{11}$, 4825 (1975).
4. E. A. Stern, D. E. Sayers and F. W. Lytle, Phys. Rev. B $\underline{11}$, 4836 (1975).
5. G. S. Knapp, H. Chen and T. E. Klippert, Rev. Sci. Instru. $\underline{49}$, 1658 (1978).
6. P. Georgopoulos and G. S. Knapp (to be published in J. Appl. Crystal. 1980).
7. G. G. Cohen, D. A. Fischer, J. Colbert and N. J. Shevchik, Rev. Sci. Instrum. $\underline{51}$(3), 273 (1980).
8. Y. Cauchois and C. Bonnelle in Atomic Inner Shell Processes II, ed. B. Crasseman (Academic Press, New York, 1975) pp. 83-121.
9. B. E. Warren in X-ray Diffraction (Addison-Wesley, Reading, MA, 1969) p. 46.
10. E. A. Stern and S. M. Heald, Rev. Sci. Instru. $\underline{50}$, 1579 (1979).
11. A.J.P.L. Polycarpo, M.A.F. Alves, M.C.M. Dos Santos and M.J.T. Carvalho, Nucl. Instr. Methods $\underline{102}$ 337 (1972).
12. J. Jaklevic, J. A. Kirby, M. P. Klein, A. S. Robertson, G. S. Brown and P. Eisenberger, Sol. St. Commun. $\underline{23}$, 679 (1977).
13. J. A. DelCuerto and N. J. Shevchik, J. Phys. C $\underline{9}$, 4357 (1978).
14. G. S. Knapp, R. T. Kampwirth, P. Georgopoulos and B. S. Brown in Proc. of the Conf. on Superconductivty in d- and f-band Metals, (Academic Press, New York, 1980) pp. 363-368.
15. D. P. Karim, P. Georgopoulos and G. S. Knapp, to be published in the J. Nucl. Technology, 1980.

EDITOR'S NOTES - CHAPTER 1

z. Chapter 11.

y. Equation (3) is valid only if the monochromating crystal has a rocking curve with an angular acceptance $\Delta\theta_c \geq \frac{W_s}{4r \sin \theta}$.
For a perfect crystal this is usually <u>not</u> the case and $\Delta\theta_c < \frac{W_s}{4r \sin \theta}$. In such a case Eq. (3) should be replaced by

$$\Delta\theta = \frac{W_a}{4r \sin \theta} + \Delta\theta_c.$$

A more detailed discussion is presented in Chapter 3.

x. Further detailed discussions are presented in Chapters 3 and 9.

w. More accurate and detailed discussions of the causes of the broadening of the energy resolution of laboratory EXAFS arrangements are presented in Chapter 3 and Lu and Stern Chapter 9.

v. Further detailed discussions are presented in Chapters 4 and 10.

u. Further detailed discussions are presented in Chapters 5 and 11.

t. See also Chapter 4.

s. As pointed out in Chapter 4, the noise level of ionization detectors is equivalent to typically 10^3 c.p.s. so that the lower intensity limit of such detectors is not as pessimistic as implied here but more like 3×10^3 c.p.s.

AN X-RAY SOURCE FOR A LABORATORY EXAFS FACILITY

G.R. Fisher
Marconi Avionics Limited - Neutron Division
Borehamwood, England

1. Introduction

The major problem in EXAFS, as far as x-ray source design is concerned is the supression of harmonics which may be diffracted by the spectrometer analyser. Until a convenient method eliminating them in the data collection system has been devised we must avoid generating them in the first place. This can best be achieved by limiting the maximum operating voltage of the tube to the 10-20kV region. There are, however, a number of unfortunate side effects associated with this action. The intensity of the Bremstrahlung, integrated over the whole spectrum is given with fair accuracy by

$$I_{TOT} = k\, i\, Z\, V^2 \qquad (1)$$

where k is a constant $(1.5 \pm 0.3) \times 10^{-9}$ related to the Bremstrahlung cross-section, i is the tube current, Z is the atomic number of the target and V is the tube operating voltage.

The efficiency of x-ray generation can be defined as

$$E = \frac{\text{Rate of emmission of x-ray energy}}{\text{Power in electron beam}} \qquad (2)$$

$$= k\, Z\, V$$

Efficiency is thus proportional to atomic number and operating voltage. This simply illustrates the necessity for using target materials with high atomic number and the disadvantage of using low operating voltages. With sealed tubes the maximum flux available is rather low but results can be obtained by careful design of the x-ray optics and spectrometer [1,2]. The use of a rotating anode source which can operate at higher powers with up to 300mA tube current in the voltage ranges of interest provides up to 20 times the useful flux obtained from sealed tube systems.[3] This allows the EXAFS spectra in the best cases to be obtained in just a few minutes. For example, Cohen, Fischer, Colbert and Sheuchik reported spectra obtained in 20 minutes and 5 minutes working at only 20kV 80mA and 20kV 40mA respectively for $(FeTPP)_2O$ and Cu^2.

The rest of this discussion will refer to some of the characteristics of a rotating anode (R.A.) source which has been developed by our engineering laboratory and is now marketed as the GX21 R.A. system. A schematic layout is shown in Figure 1 which indicates the main features of the GX21. The system is basically a 'head' (containing the anode and cathode assemblies) with full supporting equipment. The head is evacuated by turbo-molecular pumping system and suitable seals are provided which allow the anode to be water cooled and rotate at 3000 or 6000 r.p.m. The system is designed to provide spot or line focus by rotation of the head assembly about the cathode axis. The 0.5mm x 10mm focal spot in the line configuration is suitable for EXAFS analysis.

The tube voltage and filament current are stabilised by thyristor control to better than 0.1%. Electromagnetic shutters are controlled remotely from outside an optional 'see through' radiation shield. The entire system is fully interlocked.

2. Cathode Characteristics

In the majority of crystallographic applications, the x-ray source is required to operate in the region of 40-50kV with as much current as possible. A number of generators are available depending on the precise nature of the work being undertaken. In EXAFS analysis however, the requirement for low operating voltages in the 10-20kV region has some serious repercussions in the cathode design. Generally the electron gun design for x-ray sources is of the temperature limited type and the tube current is controlled completely by the temperature of the filament. Stability is achieved by a simple servo loop which monitors the tube current and controls the filament supply accordingly. At low operating voltages, the field at the filament is reduced severely by the space charge in electron beam and this results in a reduced tube current. Increasing the filament temperature has no effect and the cathode is said to be space charge limited. One solution to this problem might be to employ a Pierce gun[4]. Such devices are designed specifically for operation under space charge limited conditions and low voltages and high currents. The rate of tungsten deposition onto the target could also be very low in a good design. In practice, however, in the range of powers in which we are concerned these guns tend to be large and cumbersome and tend to generate a lot of heat which makes them inefficient. Tantalum cathodes offer some improvement but these suffer rapid degradation from small amounts of contamination which makes handling a problem. Exercising suitable care in the design of our temperature limited guns some compromise has been achieved and a generally more reliable cathode assembly has been obtained.

Taking advantage of the lower voltage hold off requirement for EXAFS experiments the cathode can be moved closer to the anode which helps to offset the space charge limiting phenomemon. Other parameters such as the depth of the filament in the focus cup and the bias voltage can also be varied to provide operating characteristics more suited to low voltage operation. The bias voltage is applied between the filament and the focus cup and is generally expressed as a percentage of the maximum value which is 500V, (i.e. 20% bias = 100V). Figure 2 shows two voltage-current characteristics for the GX21 in our laboratory set up with a 0.5mm x 10mm focal spot. Curve (A) shows the results with a cathode set for normal operation at voltages up to 60kV. Curve (B) shows the results when the cathode has been set up to operate at 15kV 300mA. Under these conditions the spot can be viewed at $6°$ to produce an effective line breadth of 0.05mm. If a 1.0mm x 10mm focus is used the space charge limit can be offset still further but of course the take off angle must be reduced to $3°$ which at low energies tends to increase the effect of absorption of x-rays into the target.

At near-space-charge-limited operating points the filament reaches its maximum temperature (about $2400°C$) and the general failure mode for filaments is by slow evaporation of the tungsten wire. It is interesting to note that at higher operating voltages the most common failure mode is by ion bombardment derived from ionisation of residual gases in the tube head. The difference is caused by the shielding effect of the space charge cloud which protects the filament.

3. Focal Spot Distribution

Although not directly relevant to EXAFS experiments it is perhaps interesting to briefly investigate the manner in which the electron beam focus is formed. As described in the previous section, a number of parameters can be varied by the user to facilitate special cathode characteristics if the standard setting is found to be unsuitable. Figure 3 shows typical electron trajectories from the 'filament-in-slot' type electron gun. It is immediately obvious that the focal spot size and shape will change depending on the distance along the beam axis Z at which the anode surface is placed. Figure 4 shows how the distribution changes. Position 'B' shows a typical distribution for a well focussed gun. In practice the change from A to C will correspond to anode cathode spacings from about 6mm to 11mm. A more sensitive parameter is the depth of the filament in the focus cup. The distributions in Figure 5 correspond to a change in depth of only about 1mm. The actual depth depends on the bias and anode cathode spacings. An increase in the bias level has a similar effect to setting the filament deeper in the focus cup. In each case, a high bias level, a large anode cathode spacing or a deep filament setting, leads to an earlier onset of space charge limiting. These points should be borne in mind when optimising for special applications.

4. Anode Design

EXAFS analysis systems employing a source like the GX21 do not operate near the maximum power for the anode. This has two important advantages. Firstly the anode lifetime between repolishing can be very long indeed, secondly the anode speed can be quite low and thus extend seal life. Specifically the GX21 anode is rated at 15kW into copper, tungsten and molybdenum targets and 11kW in gold targets. The maximum power likely to be used in EXAFS is about 6kW (20kV, 300mA). Under these conditions the rotation speed can be set at 3000 r.p.m. instead of 6000 r.p.m. resulting in a seal life in excess of 1000 hours.

Gold and tungsten targets are available among the high atomic number elements. The gold target is plated onto a copper substrate. The plating must be sufficiently thick to allow for some diffusion of the gold into the copper. Experience has shown this solution to be superior to the possibility of plating a diffusion barrier between the two layers. This is because the oscillatory stresses across the interfaces, as the periodic heat distribution flows through the anode shell, tend to break down adhesion between the various layers of material. In practice the rate of diffusion of the gold into the copper is sufficiently slow that anode lifetimes are very long indeed.

The tungsten anode has the advantage of not being affected by tungsten contamination from the cathode! It is constructed by spinning a thin tungsten shell and electroplating copper onto the inside to form a strong high conductivity composite. Naturally the anode must be carefully balanced. Our specification requires that the spatial stability of the focal spot should be better than $10\mu m$. To achieve this it is necessary to balance the anode so that the centre of mass of the anode does not deviate from the centre of rotation by more than $1\mu m$. In practice, with the anode assembled and the machine running measurements carried out on our laboratory system show the spatial stability (wideband) to be better than $5\mu m$ ($\pm 2.5\mu m$) even over long periods providing the power level is kept constant. This is partly aided by the fact that our water supply temperature never varies by more than a couple of degrees or so. The bearing life is considered terminated when the spatial stability, at maximum anode speed falls outside the specification. After many years of development maximum lifetimes have been achieved using a pair of angular contact bearings. Bearing life is often in excess of 10,000 hrs.

5. Other Features

Turbomolecular pumping has been shown to aid a superior performance from rotating anode sources. The vacuum is generally cleaner which aids component lifetimes and when service is necessary, down times are generally shorter.

The tube head on the GX21 is supported by a pillar which can be manufactured to any height. Thus the beam height above the work-top can be specified by the customer to match the particular spectrometer or goniometer being used. The electronic control circuitry is such that a simple modification can provide a facility for using the signal from an x-ray detector to control the tube current. In this way the increases in intensity at characteristic wavelengths can be smoothed out. This has been demonstrated by Knapp and co-workers with a circuit of their own design.

6. Conclusions

The higher powers available from the GX21, even at low operating voltages, together with it's adaptability to individual requirements for focus size, beam height etc... make this system an extremely useful x-ray source for EXAFS analysis. The design is based on proven and tried engineering methods which lead to a higher reliability. This coupled with a full appreciation for the implications of health and safety regulations has resulted in an overall system which is now both reliable, safe and easy to operate.

References

1   Knapp G.S., Chen H & Klippert T.E., Rev. Sci Inst $\underline{49}$ p 1658 - 1666 (1978)
2   Cohen G.G., Fischer D.A., Colbert J, Sheuchik N.J., Rev. Sci. Inst. $\underline{51}$ Pt 3 p 273 - 277 (1980)
3   Georgopoulus P and Knapp G.S. submitted to J. App. Cryst. January 1980
4   Pierce J.R. 'Theory and Design of Electron Beams' (New York - Van Nostrand) (1949)

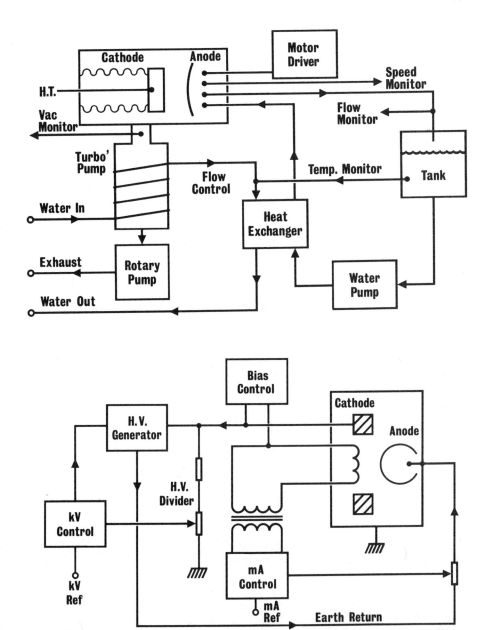

FIG. 1 SCHEMATIC LAYOUT OF GX21 ROTATING ANODE X-RAY SOURCE

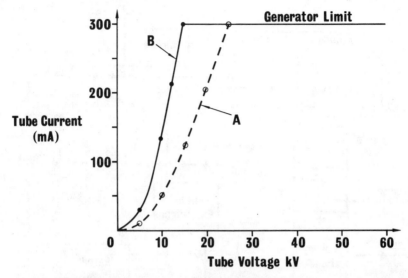

FIG. 2 VOLTAGE CURRENT CHARACTERISTICS FOR GX21 CATHODE

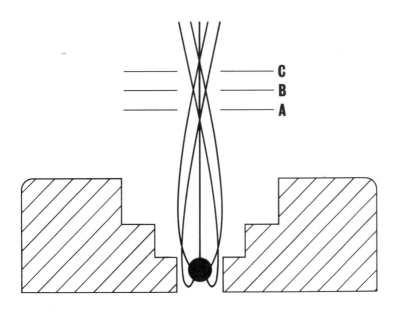

FIG. 3   A TYPICAL ELECTRON TRAJECTORY CONFIGURATION

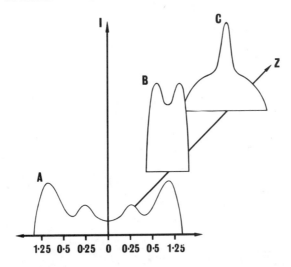

FIG. 4   VARIATION OF ELECTRON DISTRIBUTION ALONG BEAM AXIS Z

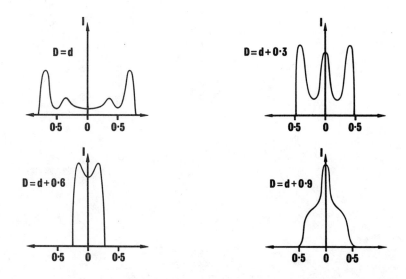

FIG. 5 VARIATION OF ELECTRON DISTRIBUTION WITH FILAMENT DEPTH

EVALUATION OF FOCUSING MONOCHROMATORS FOR AN EXAFS SPECTROMETER*

S. M. Heald

Brookhaven National Laboratory, Upton, NY 11973

ABSTRACT

Various focusing monochromators are compared and evaluated for use in an EXAFS spectrometer. For a line focus the Johann and Johansson cases are found to be most suitable. Obtaining a point focus is more difficult, with the best case appearing to be a singly bent mirror used in conjunction with the Johann or Johansson crystals.

INTRODUCTION

A typical EXAFS experiment requires $10^6 - 10^8$ counts in each of ~200 channels. If laboratory experiments are to be completed in reasonable time spans efficient focusing monochromators are necessary. These must collect the maximum solid angle consistent with an energy resolution $\Delta E/E \lesssim 5 \times 10^{-4}$, and be easily tunable.

In this paper various focusing arrangements are examined and evaluated as to their suitability for EXAFS measurements. Most of the focusing arrangements present here have been reviewed previously,[1-3] but not specifically for EXAFS uses. Also, a Monte Carlo raytracing program originally developed for synchrotron x-ray optics has been applied to the focusing monochromators to estimate their resolution and relative intensities for specific cases. The program is particularly helpful in allowing various "non-ideal" geometries to be quickly evaluated.

GENERAL CONSIDERATIONS

In general, the x-rays striking a monochromator have a spread of angles $\Delta\theta_g$. The energy resolution is determined by the convolution of $\Delta\theta_g$ with the crystal rocking curve. For perfect crystals such as Si or Ge the rocking curve is nearly rectangular with width $\Delta\theta_c$ and peak reflectivities near one. For maximum efficiency $\Delta\theta_c$ should approximately equal $\Delta\theta_g$.

For perfect crystals $\Delta\theta_c$ is determined by the choice of crystal reflection. To change $\Delta\theta_c$ for a given reflection the crystal can be made imperfect by suitable surface treatment, or an asymmetric cut can be used. The angular acceptance of a perfect crystal with an asymmetric cut[4] is

$$\Delta\theta_a \simeq \Delta\theta_c \sqrt{\frac{\sin(\theta_B+\alpha)}{\sin(\theta_B-\alpha)}} \qquad (1)$$

---

*Work performed under the auspices of the U.S. Department of Energy.

where α is positive when the incident beam makes an angle smaller than θ_B with respect to the surface. Both approaches have some disadvantages. Imperfect crystals do not have peak reflectivities as large as perfect crystals, reducing the efficiency of the monochromator over the case where a perfect crystal is used to match $\Delta\theta_g$. To significantly increase the acceptance of an asymmetric crystal α must be quite close to θ_B which is possible only for a limited energy range.

The geometrical angular spread $\Delta\theta_g$ has three major contributions. The first comes from aberrations in the focusing designs. This depends on the particular design used and will be discussed in detail in later sections. The other two are illustrated in Fig. 1 for a flat crystal. The finite source size in (a) means that each point on the crystal sees a spread S/ℓ where $S=S_o \sin(\beta)$ and ℓ is the distance from source to crystal. An important point here is that the take off angle β depends on the part of the crystal being considered. If the horizontal acceptance W_h is large then the extreme rays can see a source width significantly different from that seen by the central ray. In (b) the effect of a finite vertical acceptance W_v is illustrated. In this case the extreme rays striking the crystal out of the horizontal plane make a smaller angle with the surface than the central ray by an amount $W_v^2/(8 \tan\theta)$.

SINGLY BENT CRYSTALS

By far the best singly bent monochromators from the standpoint of easy tunability are the Rowland circle cases shown in Fig. 2, the Johann[5] and Johansson[6,7] crystals. Other designs such as the logarithmic spiral monochromator are difficult to scan in energy.

(a) HORIZONTAL

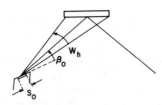

(b) VERTICAL

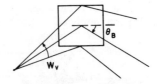

Fig. 1. Two sources of angular spread. a) Finite source size; note that S can vary if W_h is large. b) Finite vertical collection angle W_v.

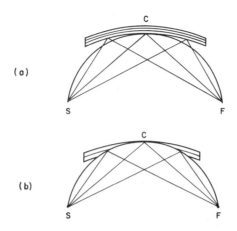

Fig. 2. Two Rowland circle geometries. a) Johann in which the crystal is bent to a radius 2R. b) Johansson in which the surface of the Johann crystal is ground to the Rowland circle radius R.

For a point source the Johansson case is an exact solution in the plane. For an actual case the only geometrical contribution to $\Delta\theta$ comes from a finite source width and the vertical collection angle as shown in Fig. 1. For the maximum intensity consistent with a given $\Delta\theta$ these two contributions should be approximately equal, although practical constraints on the size of the crystal and height of the focal line often limit the vertical collection. The horizontal collection angle can be as large as possible consistent with the practical constraints on crystal size, and the increase of the apparent source width shown in Fig. 1.

For the Johann case there is an additional geometrical contribution from the horizontal collection angle given by $W_h^2 \cot\theta/8$. There is also an increase in the focal line width although for EXAFS measurements this is usually not important. Table I compares the Johann and Johansson monochromators with a flat crystal for various cases assuming $S/2R = 5 \times 10^{-5}$. For the case shown geometrical effects dominate and the monochromators reflect only about 10% of the photons incident within the 5 eV bandwidth. Improvement can be gained by increasing the rocking curve of the crystal. One way to do this is to use Ge (400) for which the diffraction width is $\sim$2 1/2 times wider.

One might also ask if using an asymmetric cut crystal can improve the efficiency. This turns out to be not the case. For example, the source divergence can be reduced by making SC larger than CF in Fig. 2 which would increase the efficiency. However, the appropriate asymmetric crystal would have a reduced acceptance as seen in Eq. (1), and the two affects nearly cancel.

Table I Comparison of Johann and Johansson monochromators for Si (400) at 10 keV. S/2R is assumed to be 5 x 10^{-5} and the numbers are for an energy resolution of 5 eV. W_h = 4° for the Johansson case assumes R = 1 m. and a maximum crystal length of 8 cm.

	W_h	W_v	I
Flat	.0093°	2.9°	1
Johann	1.5°	2.9°	158
Johansson	4°	3.2°	465

POINT FOCUS

In many cases it is desirable to concentrate the x-rays to a small area. For the cases in Table I if R = 1 m. and the source is 10 mm high, then the focal line is ~110 mm high. Often it is difficult to make a uniform sample of this height, and make use of all the available x-rays. To reduce the height R can be reduced, but then the source width must be reduced to maintain the resolution.

Doubly bent crystals can be used to focus a point source to a point focus. They are achieved by rotating the geometries in Fig. 2 about the line SF. Again the Johansson case is a perfect solution, and in this case the contribution from the vertical divergence is also eliminated. $\Delta\theta_g$ is determined completely by the finite source size.

There are, however, two problems associated with this approach. First it is difficult to bend suitable crystals to the two radii without breakage. This can be solved by using the approximate geometry of Berreman et al.[8] with quartz. It consists of first grinding the crystal into a cylindrical shape, pressing it flat, and then bending it into a cylindrical shape in the orthogonal direction. The crystal planes are then in a doubly bent configuration with radii as small as 1000 times the crystal thickness.

The second problem is more serious. The radius of bend about SF changes with Bragg angle, and if the crystal radii are fixed the focusing condition is lost when the energy is scanned. To look at this problem a raytracing study was made of a Berreman monochromator as a function of energy. These results are summarized in Table II and Fig. 3. Figure 3 shows the change in the focal spot size and Table II also gives the characteristics of the doubly bent Johansson geometry. The crystal was assumed to be SiO_2 (203) which has nearly the same characteristics as the Si (400) used in Table I. The source spot was 0.1 x 0.1 mm (0.1 x 2 mm at 3° take off angle) which means R = 1 m for S/2R = 5 x 10^{-5}. The Berreman crystal is about the same quality of approximation to a doubly bent Johansson as is the Johann for the singly bent case. The Berreman cylinder was oriented along SF which the raytracing showed to be significantly better than the opposite case. The limit of 4° was chosen from a consideration of maximum crystal sizes and W_h = 0.6 for the Berreman case was chosen to match the Johansson ΔE.

Table II Comparison of the Berreman and Johansson doubly bent monochromators at 10 keV for SiO_2 (203). Also, given is the Berreman 10 keV monochromator at 9.5 and 10.5 keV. The source size is 0.1 x 0.1 mm, and R=1 m. The 4° limit is not due to aberrations but is estimated to be a practical limit. See Fig. 3 for the actual shape of the focal spot for the Berreman cases.

	E (keV)	W_h	W_v	ΔE (eV)	Focal Spot (mm)
Johansson	10	4°	4°	3.0	0.1x0.1
Berreman	10	.6	4	3.0	0.1x0.1
Berreman	10.5	.6	4	4.8	0.4x12
Berreman	9.5	.6	4	4.7	0.5x14

As seen in Table II scanning the Berreman monochromator does not seriously change its energy resolution, but does greatly increase the size of the focus. This could cause difficulties if the EXAFS sample is not perfectly uniform, and also destroys the major advantage of a point focus. For a truly tunable point focus monochromator the radius must be adjusted as θ is changed.

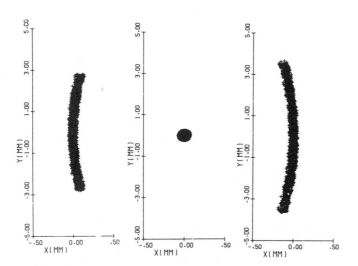

Fig. 3. Focal spot of a Berreman 10 keV monochromator as a function of energy. For left to right the energies are 9.5, 10, and 10.5 keV. The heights are half those listed in Table II since W_v = 2° rather than the 4° used there.

TWO ELEMENT MONOCHROMATORS

A point focus can also be obtained using two singly bent elements[1,9] as in Fig. 4. In this case the radii can remain fixed, but scanning in energy is very difficult since the two crystals must be kept accurately oriented with respect to each other. For $\Delta E = 5$ eV the tolerances in θ are a few seconds. The situation is simpler if one of the elements is a mirror. Then the focussing of one element is independent of energy and a Rowland circle type scan can be used for the crystal element.

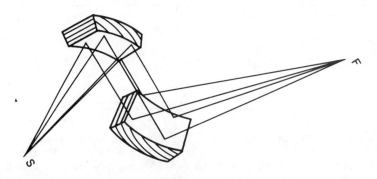

Fig. 4. Use of two crystals to obtain a point focus.

X-ray mirrors require glancing angles ϕ; the critical angle above which reflection no longer occurs is ~0.5° for a Pt surface at 10 keV. Some examples of critical angles for other cases are given in Table III. This restricts the collection aperture of the mirror. For example, the cases in Table I have a source to crystal distance of 460 mm for R = 500 mm. If a 200 mm mirror were placed midway it could collect an angle of ~θ_c or 0.5° for a 10 keV cutoff and 1° for a 5 keV cutoff (assuming the source is now a point). This is still respectable particularly since the full vertical acceptance in Table I will not always be usable.

Table III Examples of critical angles of reflection for various cases.

Element	4 keV	8 keV	16 keV
Si	0.45°	0.22°	0.11°
Ni	0.86	0.22	0.11
Pt	1.20	0.60	0.30

For a point focus the mirror should have an elliptical profile. This is quite easy to obtain by bending an appropriately shaped flat mirror.[10] There is still a difficulty in scanning energy. Since the crystal to source distance is changing, the focus will not remain on the Rowland circle. However, in some cases the motion is small and the image on the Rowland circle would not vary drastically during the scan. Such a spectrometer should be much easier to build than the corresponding double crystal case.

If the smallest point focus is not needed then an even simpler arrangement would be to use a parabolic mirror. This would collect the vertical divergence and make the beam parallel. The height of the beam is just the height intercepted by the mirror, which if the mirror is close to the course can be substantially smaller than the height of an unfocussed beam at the focus, as much as a factor of 5-10 for the same W_v. Since the final beam is parallel the mirror can be at a fixed distance from the source, and a scan carried out just as for a Rowland circle instrument with the source offset slightly above and inside the circle to account for the mirror reflection.

SUMMARY

The most crucial restriction on an EXAFS spectrometer is the necessity to easily scan wide ranges of energy. For a line focus a Rowland circle spectrometer using a Johann or Johansson crystal is most convenient. Large solid angles can be collected, and a scan is easily carried out. For maximum efficiency the crystal $\Delta\theta_c$ should be matched as much as possible to the geometrical $\Delta\theta_g$ of the spectrometer.

A point focus is more difficult to achieve. Doubly bent crystals cannot maintain a focus over a range of energy unless one radii is varied. This was demonstrated by a study of the Berreman monochromator geometry. Double crystal devices are another possibility, but are very difficult to scan since the rays must make the same angle with each crystal to within the crystal rocking curve. Probably the best candidate is a mirror-crystal combination used in a Rowland circle instrument. However, its use is probably limited to energies below about 10 keV by the extreme glancing angles required for the mirror.

REFERENCES

1. J. Witz, Act. Cryst. A25, 30 (1969).
2. B. W. Roberts and W. Parrish, International Tables for X-Ray Crystallography, Vol. III (Kynock Press, Birmingham, 1962).
3. A. Sandström, Handbuch der Physik, 30 (Springer, Berlin, 1957), p. 94-124.
4. K. Kohra, M. Ando, T. Matsushita, and H. Hashizume, Nucl. Instr. and Methods 152, 161 (1978).
5. H. H. Johann, Z. Phys. 69, 185 (1931).
6. J. W. M. Dumond and H. A. Kirkpatrick, Rev. Sci. Inst. 1, 88 (1930).
7. T. Johansson, Z. Phys. 82 507 (1933).
8. D. W. Berreman, J. Stamatoff, and S. S. Kennedy, Applied Optics

16 2081 (1977).
9. T. C. Furnas, Rev. Sci. Inst. 28, 1042 (1957).
10. J. H. Underwood and D. Turner, Proc. SPIE, 106, 125 (1977).

EDITOR'S NOTES - CHAPTER 3

z. In this chapter the implicit assumption is made that reflection occurs only at the crystal surface. Relaxing this assumption, there is a fourth contribution to the angular spread in bent crystals which is caused by the x-rays being able to penetrate the crystal before reflection. This bending aberration is mentioned in Chapter 1 and in Lu and Stern Chapter 9. For a more detailed discussion of the bending aberration refer to reference 3 in Lu and Stern Chapter 9.

DETECTORS FOR LABORATORY EXAFS FACILITIES

Dr. Edward A. Stern
Physics Department FM-15
University of Washington, Seattle, WA 98195

ABSTRACT

The measurement requirements for EXAFS are reviewed. Based on these requirements, various detectors are compared and evaluated for use in laboratory EXAFS facilities. Gas proportional counters and gas ionization chambers are recommended for transmission EXAFS measurements while for fluorescence measurements the gas scintillation proportional counter is best with the gas proportional counter and NaI(Tl) scintillation counter close seconds.

INTRODUCTION

The detection of EXAFS requires the determination of the x-ray absorption coefficient μ as a function of energy on the high energy side of absorption edges. The relation between μ, the x-ray intensity incident on the sample I_o, the x-ray intensity transmitted through the sample I, and the sample thickness x is

$$\mu x = \ln \frac{I_o}{I} \qquad (1)$$

Experimentally it is necessary to monitor I_o and I or quantities proportional to them.

This monitoring can be done directly by measuring the EXAFS in transmission or indirectly by detecting the decay of the atoms excited when the x-rays are absorbed. The transmission mode is appropriate for concentrated samples while the indirect methods are best for dilute samples. The indirect methods can monitor either the fluorescence radiation or electrons emitted from the excited atoms.

STATISTICAL CONSIDERATIONS

The statistical errors in an absorption measurement in transmission are discussed in detail by Rose and Shapiro[1] for a variety of cases. There are two primary considerations for an EXAFS measurement which requires high accuracy in determining $\mu x = \ln(I_o/I)$. The first is the choice of an optimum μx and the second is the apportionment of time in counting I_o and I, or the fraction of I_o absorbed by a partially transparent detector in front of the sample which permits a simultaneous monitoring of I and I_o. These are related, and by a simple application of counting statistics it is shown that the optimum μx is 2.6 with about 22% of the signal spent in counting I_o.[1] This result ignores the background contribution in the detector when the beam is off which is a good approximation for most EXAFS measurements.

However, other considerations make it desirable to use a smaller μx. These are in the general category of "thickness effects".[2,3] These come about when some fraction of the x-rays can pass through the sample relatively unaffected by a change in μx. This could be due to cracks or holes in the sample, harmonics in the x-ray beam, or long tails on the monochromator transmission function. Because of these, any structure in μx tends to be washed out, and the net effect is to <u>reduce</u> the amplitude of the EXAFS. The term "thickness effects" is used because the effect depends on the thickness or μx of the sample, and becomes negligible if the sample is thin enough.

To minimize the thickness effect, a uniform sample is essential. Once this is achieved the most important factor for EXAFS is generally the harmonics present in the monochromator transmission function. If μx=1.5, 5% harmonics would reduce the EXAFS by ∼10% for a concentrated sample. Harmonics can be eliminated by keeping the excitation voltage on the x-ray generator below the value to excite the lowest harmonic or by using pulse height discrimination. If, in spite of these precautions, there is still reason to suspect a thickness effect, then the only sure procedure is to measure samples of varying thickness to determine the μx below which the thickness effect is no longer important. For near edge measurements, thickness effects can occur[4] even for uniform samples and with no harmonics present because of the tail in the monochromator transmission function. Again, the only secure way to eliminate this effect is to measure the thickness dependence and find the regime where decreasing thickness no longer causes changes.

For dilute samples the change in adsorption $\Delta\mu x$ in going through an absorption edge is small and the thickness effects become less important. For these cases it is generally possible to use a total μx close to the optimum value of 2.6.

From statistical considerations the optimum fraction for I_o can be calculated.[1] For a sample thickness μx the I_o chamber should absorb a fraction.

$$f = \frac{1}{1+e^{\mu x/2}} \qquad (2)$$

to minimize statistical errors. For typical values of μx, f is 20 - 40%. Such an I_o monitor is useful for all types of EXAFS measurements.

As mentioned previously, the fluorescence detection method becomes superior for dilute samples.[5] The accuracy of the fluorescence and transmission techniques can be compared by determining the noise in measuring μ_s, the adsorption edge contribution of the element of interest. It can be shown that the two techniques have equal noise for dilute samples when

$$\mu_s x \approx \left| \frac{2\alpha e^{\mu_b x}}{(\mu' + \mu_b) x} \right| \qquad (3)$$

where μ_b is the absorption coefficient of the rest of the sample, μ' is the adsorption coefficient in the sample at the fluorescent energy and α is the fraction of the excited atoms whose presence is detected by a fluorescent x-ray. For the Fe K-edge the fluorescence yield is 0.3, and a typical x-ray filter detector[5] (see below) might collect $\sim$10% of 4π and attenuate the fluorescence signal by a factor of 2. This gives $\alpha=0.015$ and if $\mu_b x=2.6$ and assuming $\mu'=\mu_b$ then the two techniques give equal signal to noise ratios at $\mu_s/\mu_b=.03$. Thus the fluorescence technique becomes advantageous in terms of counting statistics when $\mu_s/\mu_b \leq 0.03$ or when the absorption edge step due to the element of interest $\mu_s x \leq 0.08$ in a sample of optimal thickness $\mu_b x=2.6$.

There is only one technique that can attain the large α assumed above so as to make fluorescent advantageous at such a high concentration and give correspondingly large counting rates, namely the x-ray filter assembly. The other fluorescent detectors such as solid state detectors and crystal monochromators[7] subtend an order of magnitude smaller solid angle and these are not advantageous till dilutenesses at least an order of magnitude less than $\mu_s x=0.08$. The counting rate from a laboratory EXAFS appratus for such dilute samples will be quite low making the applicability of the other fluorescent detectors marginal.

X-ray filters use x-ray absorption edges to selectively attenuate the scattered radiation. This is possible since the background radiation, which consists of elastically and Compton scattered incident radiation, is of higher energy than the fluorescent signal. By choosing the appropriate filter, the absorption edge can be placed between the background and fluorescent energies preferentially absorbing the former.

The main difficulty with filters is that radiation absorbed in them is re-emitted as fluorescent radiation which is relatively unattenuated by the filter. However, a suitable set of Sollertype slits can block most of this filter fluorescence and allow the ideal filter characteristics to be approached[6]. With filters, a large collection efficiency is possible, and changing the energy setting is simply accomplished by changing filters. Although the background discrimination is not perfect, for many applications this is more than offset by the large collection efficiency.

TYPES OF DETECTORS

The ideal detector for a laboratory EXAFS facility should have the following characteristics: (a) It should be highly linear over a dynamic range of 100 to 1 so as not to introduce "glitches" as an emission line is scanned through. (b) It should have a high and variable efficiency over the x-ray energy range of 2.5-20 KeV. The variable feature is for use as a partially transparent I_0 detector with the capability to optimize the f of Eq. (2). (c) It should be able to handle intensities up to 10^7 photons per second. (d) It should have good energy resolution. The good resolution is necessary for fluorescence use where there is the need to discriminate between the fluorescent radiation and the elastic background, typically a 15% resolution requirement. The detector also needs to discriminate against the higher harmonics, though this is not as strict a resolution requirement. (e) Finally, for fluorescent use the detector should have a large area to maximize the counting rate.

A. PULSE

Detectors can be divided into two categories, pulse and d.c.. In the pulse type each photon is individually detected by the pulse it produces while in the d.c. type only the integrated number of photons is detected by measuring a current.

In this paper there is no attempt to describe in any detail the basic physics of operation of the various detectors. This information is available in the references. Here the emphasis is on the applicability of the detectors to the specialized needs of a laboratory EXAFS facility.

An ionization chamber[8] is not useful in the pulse mode for the low energy x-rays present in EXAFS facilities. Its only application is in the d.c. mode which is discussed later.

A proportional counter[9] consists of a chamber filled with a gas and a positively charged wire that can attract electrons. X-rays in the energy range of interest for EXAFS are absorbed predominantly by the photoelectric effect. The electrons freed in the ionization produced by the photoelectron are attracted to the wire. Typically about one electron-ion pair is produced per 30eV incident x-ray energy. The electric field in the vicinity of the wire is great enough to cause a cascade process and further multiply the number of electrons freed and thus collected. This multiplication factor typically ranges from $10^2 - 10^4$.

In a scintillation counter[10] the x-ray is also absorbed by the photoelectric effect. The subsequent ionization caused by the photoelectron excites particular energy levels which decay by emitting fluorescent photons in the visible and near ultraviolet region. The fluorescent photons are detected by a photomultiplier tube producing the pulse.

More recently the gas scintillation proportional counter has been developed[11]. In this device the x-ray is absorbed in a very pure noble gas, again by the photoelectron effect. The electrons freed by the subsequent ionization are drifted between two grids in modest electric fields. The electrons are elastically scattered by the noble gas atoms till they are accelerated to sufficient energy for exciting the atoms. The electrons then inelastically scatter and the process is repeated till again the electrons have enough energy to excite another atom. Each excited atom in the pure environment de-excites essentially entirely by emitting photons. By this mechanism and a sufficiently large spacing of the accelerating grids, typically 300 photons can be emitted for each electron. The photons are detected by a photomultiplier tube. The pulse in the photomultiplier tube which is the product of the number of ionized electrons and the number of detected photons per electron is the signal that is detected.

Solid state detectors[12] consisting of Si or Ge can be used as ionization detectors when the semiconductor material is treated with appropriate impurities so as to lower its dark current (when no ionization is present). This is usually done by adding Li and then operating the material at liquid nitrogen temperatures. In this material the average energy per ion pair is about 3eV, ten times less that for has ionization detectors. The signal is 10 times larger and the statistical fluctuations are smaller permitting pulse operation in the x-ray region with good energy resolution. The signal from the detectors is amplified electronically before being recorded.

B. CURRENT or DC

Any of the pulse detectors discussed above can be used as a dc or current detector by simply measuring the average current in the device. However, since it is necessary to obtain the x-ray intensity with great accuracy, typically 0.1% or better, the stability of the detectors should be that order or better. All of the pulse detectors that employ multiplication, either by a photomultiplier tube or within the gas itself as in the proportional counter, will not inherently have the needed stability. The exponential multiplication process amplifies changes in the output of the detector due to variations in the power supply voltage. Thus a voltage stability of 10^{-4} or better may be required to obtain the output stability required, not an easy condition to attain.

The two detectors that do not employ multiplication, namely, the gas and solid state ionization detectors, have an output quite insensitive to power supply voltage when the voltage is chosen to be in the so called "plateau" region. They are therefore best suited to be operated in the current made. For a given x-ray intensity the output current from an ionization detector rises initially as the voltage across the electrodes increases and then reaches a constant value, the plateau region as indicated in Fig. 1. If the voltage is further raised, a point is reached where a new increase in the current ensues as multiplication starts. The initial rise occurs as the voltage becomes large enough to start collecting the electron-ion pairs before they can recombine and the plateau is reached when all electron-ion pairs in the sensitive volume are collected.

C. EXOTIC DETECTORS

There is always the possibility that future developments will make it clear that types of detectors not presently available can be used advantageously to detect x-rays for a laboratory EXAFS facility. A type that is presently being developed, though it does not appear at the moment to be advantageous for laboratory EXAFS use, is a liquid detector[13]. This detector can be used in both the scintillation and ionization mode. However, it does not have any obvious advantages for laboratory EXAFS use over the ones mentioned above and does have the disadvantages of the inconvenience of operation at low temperatures.

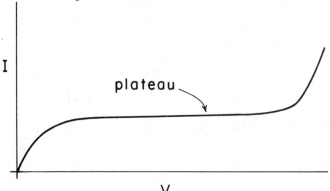

Fig. 1. A schematic of the variation of current I collected by an ionization detector as a function of voltage V applied across the electrodes for a constant flux of x-rays. The plateau region is the useful range of the detector in the ionization mode.

CHARACTERISTICS OF DETECTORS

A. PULSE

Pulse detectors have an inherent interdependence of their linearity with the counting rate. There is a minimum time between pulses required to resolve each one, called the dead time τ. Pulses at t_1 and t_2 can be resolved, i.e., counted as separate pulses, if $(t_2 - t_1) \geq \tau$. Denoting N as the true counting rate per second and N' as the measured rate, then they are related to one another, if $N\tau \ll 1$, by

$$N' \approx N(1-N\tau) \quad (4)$$

The dead time leads to a typical counting rate N' versus N as illustrated in Fig. 2. The actual counting rate N' reaches a maximum and then decreases as N increases. By measuring the curve in Fig. 2 to an accuracy of 0.1% or better[14] it would be possible to correct N' by using a computer. However, this would be practical to the required accuracy only till $N\tau \approx 0.4$, the maximum. Thus the maximum counting rate feasible from a pulse detector in use with a laboratory EXAFS facility would be

$$N_{max} \approx \frac{0.4}{\tau} \quad (5)$$

Other significant characteristics for laboratory EXAFS detectors were presented at the beginning of the previous section. The Table lists all of the significant characteristics of pulse detectors as they pertain to laboratory EXAFS facilities. The gas detectors have the additional feature of easily varying their efficiency

Table I Characteristics of various pulse detectors for laboratory EXAFS facilities

Detector	τ (seconds)	Intensity range (photons/sec)	$\frac{\Delta E}{E}$ at 10KeV	Energy Range	Remarks
Gas Proportional Counter	10^{-7}	$0-4 \times 10^6$	14%	1-30KeV	Large Area, Variable Efficiency
Solid State, Si	10^{-6}	$0-4 \times 10^5$	3%	1.7-30KeV	77K Operation, Small Area
Scintillation, NaI (Tl)	10^{-6}	$0-4 \times 10^5$	35%	1.7-30KeV	Large Area
Scintillation, Organic	0.3×10^{-8}	$0-10^8$	$\sim$50%	1.7-30KeV	Large Area
Gas Scintillation Proportional Counter	2.6×10^{-6}	$0-2 \times 10^5$	7%	1-30KeV	Large Area, Variable Efficiency

by changing gas pressure which makes them useful as partially transparent I_o monitors.

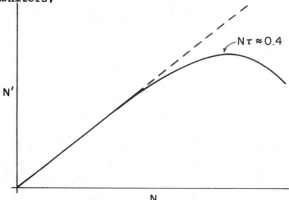

Fig. 2. A typical curve showing the relationship between the detected counting rate N' versus the actual counting rate N for a pulse detector. The dead time of the detector is τ.

B. CURRENT

As mentioned in part B of the previous section, ionization detectors without multiplication have the current stability required for EXAFS measurements. Their linearity is not limited by dead time considerations making them extremely linear to any practical intensity encountered in a laboratory EXAFS facility.

Their limitations are no energy discrimination and a minimum useful counting rate due to amplifier noise. A good current amplifier used in conjunction with an ionization detector has an equivalent current noise of

$$i_N \approx 10^{-15} \text{ amp} \tag{6}$$

It is desireable to be noise limited by statistics and not i_N in order to introduce the lowest possible noise, since one cannot do better than statistical noise. At 10 KeV a single x-ray photon in a gas produces $10^4/30 = 333$ ion pairs $= 5.3 \times 10^{-17}$ coulombs. For an x ray intensity of N photons per second at 10 KeV each, the statistical fluctuation rms current is

$$i_s^g = (5.3 \times 10^{-17}) N^{1/2} \tag{7}$$

For statistical noise to dominate let $i_s = 2i_N = 2 \times 10^{-15}$ amp. Using (7) this leads to a value of

$$N > N_m^g = 1.4 \times 10^3/\text{sec} \tag{8}$$

Thus ionization detectors will give the same statistical noise for 10 KeV photons as do pulse detectors when the intensity is greater than

1.4×10^3/sec. For other energy photons, E, the minimum counting rate for attaining statistical noise is

$$N_m^g = (\frac{10}{E})^2 \; 1.4 \times 10^3/\text{sec} \qquad (9)$$

where E is in KeV. The useful intensity range for a gas ionization detector is $N > N_m^g$. Gas ionization detectors are simple, are linear over a large dynamic range, can be made with large areas, and can be made partially transparent so that they can be used to measure I_0. Their disadvantages are no energy discrimination and they are noisier for $N < N_m$.

For solid state ionization detectors, because of the 10 times larger number of electron-ion pairs produced, the minimum useful counting rate is

$$N_m^s = (\frac{10}{E})^2 \; 14/\text{sec}, \qquad (10)$$

or a factor of 100 better than their gas counterparts. The solid state detectors, in contrast to gas ones, require cooling to liquid nitrogen temperatures to attain these characteristics, they cannot be conveniently used in the partially transparent mode to measure I_0, nor can they be easily made to cover large areas.

CONCLUSIONS

In this section, based on the characteristics presented above, recommendations are made for the best detectors for use in a laboratory EXAFS facility.

For transmission EXAFS measurements it is best to measure I_0 simultaneously with I in order to cancel out time fluctuations in the source or mechanical wobble in the apparatus which produce intensity changes with time. To do so a partially transparent detector whose efficiency can be varied to match the optimum f of Eq.(2) is required to monitor I_0. The gas proportional counter and the gas ionization chamber are best in this regard. A 100% efficient detector is best to monitor I. With $f \approx \frac{1}{2} \approx e^{-\mu x}$, I_0 can be up to 1.6×10^7/sec and still permit operation with proportional counters to measure both I_0 and I, as can be seen from the Table. The proportional counter is preferred over the ionization chamber because it can reject harmonics by energy discrimination. As pointed out by Y. Yacoby in the detector workshop, Chapter 10, this harmonic rejection permits operation of the x-ray generator tube at higher voltage with significantly higher I_0 generation. If I_0 is greater than 1.6×10^7/sec, a multiwire proportional counter can be employed with separate electronics for each wire so that each wire can handle the full counting rate given in the Table. If n wires are designed to each intercept 1/nth of the intensity, then an $I_0 = n1.6 \times 10^7$/sec can be accomodated. An ionization chamber cannot discriminate against harmonics and in order to prevent errors introduced by the harmonics it is necessary to operate the x-ray tube at excitation voltages below the harmonic generation. For this reason, gas ionization detectors normally will require x-ray tube operation at lower than optimum voltage with the attendant loss of I_o. The relative simplicity of an ionization detector still may commend its use in spite of the I_o loss.

As an example of the use of ionization chambers, a scan is shown in Fig. 3 as made on the EXAFS facility at the University of Washington which used a partially transparent ionization chamber to monitor I_o and an absorbing one to measure I. The x-ray tube employs a fixed tungsten anode which has five significant tungsten Lβ emission lines within the scan as indicated by the arrows. The numbering of the arrows corresponds to the number of the Lβ lines. The Lβ$_1$ line is about 60 times more intense than the continuum. The excitation energy of the x-ray tube is 16 KeV, below the value of any harmonic generation throughout the whole scan. Standard electronics was employed without any attempt to obtain ultra linear behavior. Yet the effects of the emission lines are almost eliminated by this simple system. In addition, the intensity I_o is about 5×10^5 so that the all counting rates are larger than $N_m^o \approx 2 \times 10^3$/sec.. The noise is still statistical in the ionization chambers.

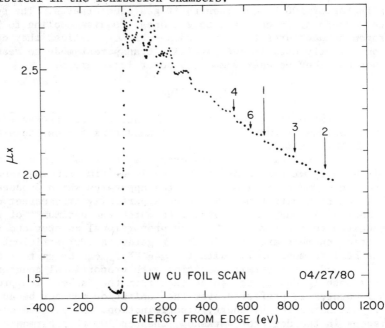

Fig. 3. A scan of μx versus x-ray photon energy for a copper foil as measured on the University of Washington EXAFS facility. Ionization chambers are used as both I_o and I detectors. The source of x-rays is a tungsten fixed anode tube. The location of five Lβ emission lines of tungsten are indicated by the arrows. The numbers by the arrows indicate the number of the Lβ line.

In measuring I it is desirable to have as high efficiency as possible with the highest possible counting rate. The plastic scintillators could be used for this purpose but their energy resolution is poor enough that a significant fraction of harmonic photons will still be detected at settings of the energy discrimination which do not lose significant numbers of the fundamental photons. The best I detector again is the gas proportional counter because by multiwire operation it can be made to handle the intensities encountered in a laboratory EXAFS facility with the required linearity and energy resolution. Next best is the gas ionization chamber because of its simplicity, though one loses I_o because of the need to operate the x-ray generator at voltages below harmonic excitation.

For fluorescence measurements, monitoring of I_o is still required and the partially transparent detector is still best as for transmission. However, because fluorescence measurements are only advantageous when the sample is dilute, the expected fluorescence counting rate will be below about 10^4/sec for laboratory EXAFS facilities.

The requirements for an ideal fluorescence detector are: (1) Large surface area to subtend as large a solid angle as possible. (2) Good energy resolution to resolve the elastically scattered radiation from the fluorescent signal. There is an 11% difference in these two energies at the Fe K-edge. (3) Ability to count 10^4/sec. or less with good linearity.

The difficult energy resolution requirement of (2) can be satisfied by using an x-ray filter[5] as discussed above and eliminating its requirement in the detector itself. The need for handling low counting rates of (3) necessitates the use of pulse detectors. From the Table and the above requirements the best detectors are:
1. Gas scintillation proportional counter
2. Gas proportional counter
3. NaI (T) scintillator

The relative advantages of the best in the above list to the worst are so marginal that other considerations such as cost, commercial availability, geometric size all could reverse the order. Any one of the above three detectors together with an x-ray filter would make a satisfactory fluorescent detector for a laboratory EXAFS facility.

REFERENCES

1. M.E. Rose and M.M. Shapiro, Phys. Rev. 74, 1853 (1948).
2. L.C. Parrott, C.F. Hempstead, and E.L. Jossem, Phys. Rev. 105, 1228 (1957).
3. S.M. Heald and E.A. Stern, Phys. Rev. B16, 5549 (1977).
4. D.M. Pease, L.V. Azaroff, C.K. Vaccaro, and W.A. Hines, Phys. Rev. B19, 1576 (1979).
5. J. Jaclevic, J.A. Kirby, M.P. Klein, A.S. Robertson, G.S. Brown and P. Eisenberger, Solid State Comm. 23, 679 (1977).
6. E.A. Stern and S.M. Heald, Rev. Sci. Instr. 50, 1579 (1979).
7. J.B. Hastings, P. Eisenberger, B. Lengeler, and M.L. Perlman, Phys. Rev. Lett. 43, 1807 (1979).
8. H.W. Fulbright, Chapter in Handbuch des Physik, vol. XLV (Springer-Verlag, Berlin, 1958) coedited by E. Creutz.
9. I. Veress and A Montvai, Nucl. Instr. and Methods (Netherlands) 156, 73 (1978).
10. W.E. Mott, Chapter in Handbuch der Physik, vol. KLV, (Springer-Verlag, Berlin, 1958) coedited by E. Creutz.
11. D.F. Andersen, T.T. Hamilton, W.H.- M. Ku, and R. Novick, Nucl. Instr. and Methods (Netherlands) 163, 125 (1979).
12. F.S. Goulding, J.M. Jaclevic, and A.C. Thompson, in "Workshop on X-Ray Instrumentation for Synchrotron Radiation Research" Stanford Synchrotron Radiation Laboratory report No. 78/04, edited by H. Winick and G. Brown (1978).
13. M. Miyajima, K. Masuda, Y. Hoshi, T. Doke, T. Takahashi, T. Hamada, S. Kubota, A. Nakamoto and E. Shibamura, Nucl. Instr. and Methods (Netherlands) 160, 239 (1979).
14. M.A. Short, Rev. Sci. Instr. 31, 618 (1960).

Chap.5

LABORATORY EXAFS FACILITY HARDWARE AND SOFTWARE

W.T. Elam
University of Washington, Physics Department, FM-15
Seattle, WA 98195

INTRODUCTION

This paper will attempt to cover the hardware related to computer control of an EXAFS experiment, the computer itself, and the computer programs unique to this experiment. Since this is one paper in a workshop proceedings, no attempt will be made to cover the hardware or electronics involved in the experimental measurement itself. Also, because of limited space and because of the vast differences in individual experiments, the paper will be mostly limited to general information. Greater detail in any area as well as hardware drawings and computer programs are available by contacting the author.

The information presented here will be divided between a description of the approach taken at the Univ. of Washington in constructing the EXAFS facility and a somewhat more idealistic approach based on the experience gained in this construction and the author's own opinions. These opinions will, hopefully, be clearly indicated as such. The approach taken in constructing our facility at the University of Washington was to make everything as inexpensive as possible. This implies that the design and construction was somewhat more difficult; its main recommendation was its success in producing a workable machine.

The author is indebted to many people, most of whom were directly involved in constructing our apparatus. Their ideas made the facility a success, and much of the information in this paper is simply a compilation of their efforts. Ed Stern conceived the design and served as the source of inspiration and decisions when everyone else was stumped. Steve Heald did most of the initial design and purchase of the computer and interface hardware. Grant Bunker, Ed Keller, and Jacob Azoulay completed the details of the design and actually constructed the computer interfaces and experimental control hardware. Lu Kunquan was responsible for most of the x-ray optics. Finally, the software was unwittingly written by Jon Kirby, whose programs for the beamlines at SSRL were shamelessly stolen and adapted to our facility. The author's contributions were to design the interfacing, lead the design and construction in the last few months and write this paper.

BASIC FUNCTIONS

A review of the principal functions of a laboratory EXAFS facility will be given first, both to familiarize the reader with the nomenclature used herein and to serve as a starting point for discussion. The apparatus must measure the x-ray absorption coefficient of a sample at a given x-ray energy, change this energy some predetermined amount, then re-measure the absorption. This cycle is repeated for many different energies and the data are stored in some form for later analysis. The measurement implies 2 input signals: one each for the x-ray flux before and after the sample. These inputs must have some form of analog-to-digital conversion and may include integration over a selected interval to achieve the desired signal-to-noise ratio. Similarly, the changes in energy require a control output - usually a stepping motor to change the Bragg angle of a crystal monochromator. More sophisticated monochromators may require more sophisticated controls. In addition, other functions may be desired, such as attenuation of the incident beam to compensate for sudden changes in source brightness (as at emission lines). As experiments become more precise and more is known about their sensitivity to various parameters, these parameters may also need to be controlled.

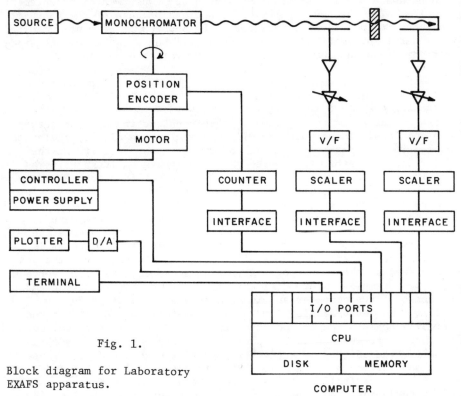

Fig. 1.

Block diagram for Laboratory EXAFS apparatus.

Figure 1 shows a block diagram of an apparatus for performing the above functions. Here the computer has control outputs and signal inputs connected to its input/output ports. The x-ray energy is changed by a stepping motor under computer control. An intermediate controller is necessary to use the logic signal from the computer to switch the high currents required by the motor. On the signal inputs, the analog-to-digital conversion scheme chosen was one employing voltage to frequency conversion. The voltage signal from the detector is amplified and sent to a V/F converter, whose output is a series of pulses whose frequency at any instant is proportional to the input voltage. Counting these pulses for a preset time gives a number proportional to the integral of the detector signal, and thus the integral of the x-ray flux. The scalers used to count the pulses have digital output signals which are sent to the computer through an appropriate interface. This method of A/D conversion was chosen because it is simple, inexpensive, and includes the integration in a straightforward manner. The data are stored in the computer's memory temporarily, and a magnetic disk (or other semi-permanent storage system) serves as longer term storage for data and programs.

Several embellishments are also shown in Fig. 1. Some form of terminal is useful to transmit information to and from the programs running the experiment. A plotter is also necessary if the data is to be examined before it is sent on to the analysis phase (or if analysis is to be performed on the same computer). A simple X-Y recorder can be used for this purpose in conjunction with a digital-to-analog converter (the reverse of the process on the signal inputs, and somewhat more easily accomplished). This type of analog output signal is useful for other functions as well, such as controlling incident beam intensity. Finally, a position encoder is shown on the monochromator drive. This serves two purposes: it makes the calibration of the monochromator immune to computer problems (either software "crashes" or hardware bugs) and it provides feedback on the monochromator motion. Neither are absolutely necessary, but they can considerably reduce the complexity of the programming required to keep the monochromator in calibration and can save a great deal of time wasted in taking worthless data (the reason for which may not be readily apparent without some way of checking the monochromator movement). It may also be convenient to have the computer program sense the conditions of the monochromator limit switches and stop taking data if they are activated.

With the hardware part of the apparatus thus outlined, it is appropriate to take a quick look at the programs which drive it. The main function of the software is to scan the x-ray absorption as a function of energy. There are also various support and auxillary functions necessary for setting up the scan and for testing the apparatus. These include manually moving the monochromator, setting up the scan parameters, such as energy range and integration time, and various initalizations, both software and hardware (most notably the monochromator calibration). These functions require manipulation of the control outputs and reading the signal inputs, which are handled in hardware driver routines in conjunction with input-output routines. A simple diagram giving a possible organization for such a system is shown in Figure 2.

HARDWARE VS. SOFTWARE

Given the basic outline for the apparatus and the programs, the next consideration is the boundary line between hardware and software: which details should be written into the computer programs and which require or are more easily handled by physical devices. Software, as its name implies, is easier to construct and much easier to change. It also involves very little direct capital investment. The average computer program, if well written, may involve a considerable development time (and expense), but even this is usually less than

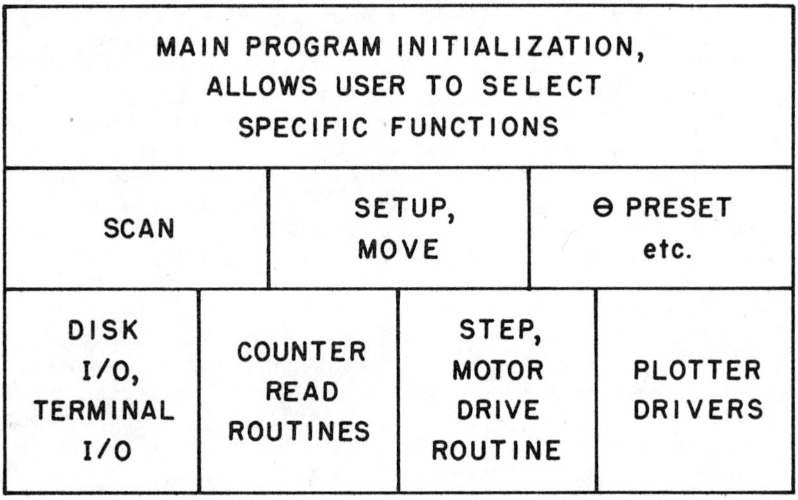

```
┌─────────────────────────────────────────────────┐
│           MAIN PROGRAM INITIALIZATION,          │
│           ALLOWS USER TO SELECT                 │
│           SPECIFIC FUNCTIONS                    │
├──────────────┬──────────────┬───────────────────┤
│              │   SETUP,     │   θ PRESET        │
│    SCAN      │   MOVE       │      etc.         │
├──────────────┼──────────────┼─────────┬─────────┤
│   DISK       │  COUNTER     │ STEP,   │         │
│   I/O,       │  READ        │ MOTOR   │ PLOTTER │
│   TERMINAL   │  ROUTINES    │ DRIVE   │ DRIVERS │
│   I/O        │              │ ROUTINE │         │
└──────────────┴──────────────┴─────────┴─────────┘
```

Fig. 2.

for a physical device of comparable complexity. Consequently, the preferred approach is to do as much as possible in the computer program and to keep the electronic hardware as simple as possible.

However, some tasks are inherently better handled by hardware. Some hardware is, of course, absolutely necessary. No amount of programming skill will allow a computer logic signal to provide the several amperes of current required by a stepping motor. In addition, the number, duration, and repetition of the pulses must be controlled. The above philosophy would decree making the motor controller a simple current switch and write the program to control the pulse count and length. However, the details of the current pulse shape sent to the motor are fairly important, and commercial units are available which control the shape of the current pulse as well as containing the switching circuitry. Such a unit alleviates the need for special, very fast programming techniques to control the pulse duration.

Fast timing is another function more appropriately or more easily handled by hardware. A more trivial example is a counter. A program could be written which continually tests an input and increments an internal variable every time a pulse appears. Such a counter costs nothing (assuming the computer is already present), but it is slow and ties up the computer completely. Fast hardware counters are cheap and readily available.

The second function appropriate for hardware is the retention of critical, volatile information. An excellent example is the monochromator position. This position could be stored in the internal computer memory. Such an approach requires special software to prevent writing over this location during normal operations or during a "crash". Also, this memory location must always be updated separately every time the monochromator is moved. This makes software changes for new experiments susceptible to subtle and disastrous errors. These problems can be avoided by the shaft encoder and counter scheme shown in Fig. 1. The information is always available by reading the counter and is updated automatically, since it reflects the actual position. This feedback on the actual position can also be very important for critical parameters in a complex experiment.

The examples given above are intended to illustrate that hardware is sufficiently superior for some tasks to justify the additional time, expense, and inflexibility which it entails. Most functions are, however, more appropriately included in the programs, i.e., in software.

COMPUTER INTERFACING

Perhaps the most difficult problem encountered in designing a computer controlled experiment is the problem of interfacing the equipment to the computer. The approach taken for our facility at the University of Washington was to construct custom interfaces to our computer for commercially available but inexpensive counters and stepping motor controllers. The major advantage of this method is its low cost; it is also customized to our particular needs. However, its uniqueness implies that little or no documentation is available on it and it is not easily fixed or replaced if a problem develops. It also involves a fairly large investment in development and construction time.

Fortunately, several alternate approaches are available. The most appropriate seems to be the CAMAC system developed for use in coupling data from high energy physics experiments to small computers for processing and storage. This system contains all of the necessary functions for running an EXAFS experiment and computer interfacing has already been done. The system is in a convenient, modular, plug-in design with its obvious advantages in repair and flexibility. Finally, the necessary driver routines to manipulate the hardware are available from several sources. This system has been in use at SSRL for several years. The advantages of having similar operating characteristics with this facility and of being able to implement software and designs there cannot be overlooked.

The main disadvantage of CAMAC equipment is its expense. To give the reader an idea of how this compares to the "low cost" approach and to the costs of the remainder of the experiment, a list of the required functions and their approximate cost is given in Table I.

No error bars are given for these prices, as they would be so large as to be meaningless. The reader is cautioned that these prices are intended only for comparison estimates. Use of these figures for budgetary planning is done entirely at your own risk. It should also be pointed out that this list includes only the computer and experimental interface gear. Computer terminals, amplifiers, shaft encoders, motor power supplies, etc. have not been included. Also, a few tips gleaned from the experiences of several people may be helpful. The CPU in the list is an LSI-11 which plugs directly into the controller slot on the CAMAC crate. This CPU has the advantage that it is compatable with the entire line of Digital Equipment Corporation minicomputers. A large amount of commercial and user

Table I

Function	Cost (1979)	
	CAMAC	U.W. Facility
Scalers	$ 500	*
θ counter	$ 700	*
Motor controller	$ 800	*
CPU	$3600	$4000
Memory	$1300	$1000
Disk	$3700	incl.
Crate / P.S.	$2000	N/A
Plotter D/A	$1400	(*)
Real-time clock	$ 625	incl.
TOTALS	$14,625	$5000

* in-house construction

not included: Terminal, Plotter, Amps or SCA's, V/F conv., DVM's, Shaft encoder, Motor power supply software exists for these machines, including that developed for SSRL. Inexpensive, "home" computers are not recommended as they do not have the software support or the accessories necessary for laboratory use. Computing power per se (CPU and memory) is relatively cheap compared to other devices, and past trends indicate that even the most exorbitant amounts soon become overloaded. A similar but more pronounced trend exists for disk storage. For this reason, the 5¼" diameter "micro-disk" systems are not recommended. Good results have been obtained with 8" floppy disk systems from Data Systems, Inc. These are available in convenient form from the makers of the CAMAC CPU and have been included in the list. Many computer installations are now beginning to install hard disks with mega-byte storage capacity.

There are, of course, many other approaches to computer interfacing. The other scheme widely used, which also contains the functions necessary for experimental measurement and control, is Nuclear Instrumentation Modules (NIM). Other methods are usually named for the computer input/output bus with which they are compatable. They include the S-100 hobbyist bus (used in the UW computer), Intel's SBC-80 bus for 8080 microprocessors, the LSI bus structure offered by DEC, and the IEEE-488 bus (also referred to as GPIB or HPIB). Unfortunately, these systems do not have the functions required for control of EXAFS experiments (most notably stepping motor controllers). The IEEE-488 bus bears watching for future developments. While present equipment for this bus is rather expensive and limited, it is a very easy to use, sophisticated, and well designed system.

COMPUTER ORGANIZATION

Before discussing the experimental programs themselves, this section will give an overview of the software organization of a

COMPUTER ORGANIZATION

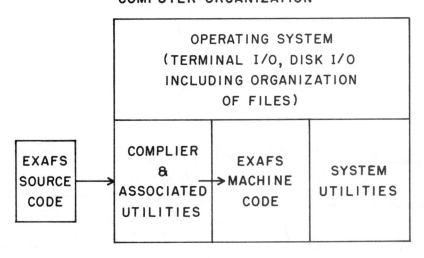

Fig. 3

typical small computer. Figure 3 is a simplified sketch illustrating this organization. An operating system, or supervisory program, is responsible for loading and running user programs upon command and for various standard utility functions such as terminal input and output and disk input and output including organizing the disk into its file structure. Some of these functions are available directly via user commands, such as listing disk files; others are called from a user program, such as creating a new file and writing information to it. Some of these functions are performed by system utilities which run as user programs but are supplied with the operating system.

The experimental programs are usually written in some convenient language (in our case, FORTRAN), and then translated into the computer's machine language by a compiler (which also runs as a user program). At the U.W. facility, the operating system is CP/M from Digital Research, Inc. and the FORTRAN compiler is supplied by Microsoft, Inc. For an LSI-11 system, Digital Equipment Corporation provides a selection of operating systems and compilers.

A possibly useful extension not found in simple operating systems is foreground/background capability. This allows two independent programs to be running on the computer, with the background program running while the foreground program waits, for instance during integration times. This is useful for plotting and analysis of previous data, but requires slightly more sophistication both in hardware and in programming techniques.

The FORTRAN language is well suited to experimental applications. It is known by most scientists to some degree and is readily available. Other possible languages are PASCAL, PL/M, and assembler. These suffer from being less well known and are less stable and may not be available for some computer systems. In addition, PL/M and assembler are more machine dependent. These factors make software in these languages more difficult to transport or exchange between various installations. BASIC language is not recommended because it is slow and many tasks necessary for experiment control are awkward to implement in BASIC. An approach often taken by programmers is to write most of a software system in FORTRAN with parts in assembly language. The author's experience is that this is necessary only in extreme cases, as carefully written FORTRAN generates code within a factor of two as efficient as assembly without special programming tricks and the FORTRAN code is more comprehensible and more easily maintained and debugged. The University of Washington software is written entirely in FORTRAN, including all hardware drivers and timing routines.

EXPERIMENTAL SOFTWARE

The functions of the programs for running the EXAFS apparatus can be divided into five major areas: 1) data taking, 2) data manipulation, such as plotting, 3) software setup, such as selecting the scan range and point spacing, 4) hardware initialization, such as monochromator calibration, and 5) a variety of "convenience" functions. The usual procedure for organizing such a set of routines is to place each function into a separate subroutine. These function subroutines then form the bulk of the software and contain the desired procedural information for achieving a particular operator identifiable task. In addition, two other sections of the program are necessary. The first is a routine which accepts the operator's input concerning which task is to be performed and which calls the appropriate subroutine. This is usually included in the main program. Secondly, a set of very low level routines will be necessary to interface between the hardware itself and the function subroutines. For example, in our system, a small routine accepts the signals from the scalers and converts them to a numeric value in some internal FORTRAN variable. The data taking routine simply calls this small routine whenever it needs a signal input. Such a separation is not absolutely necessary, but it prevents redundant code and, if properly arranged, allows hardware changes to be handled by changing only a small, low level routine. This structure was illustrated in Fig. 2. It resembles the structure of the operating system, which has proven to have many advantages. For programs which must be maintained by future generations of laboratory workers or students and which must be continually altered to meet new demands, a carefully separated hierarchal structure is mandatory.

The data taking function consists of a well defined series of steps: read the monochromator position (θ), the incident flux (I_o), and the flux transmitted through the sample (I_t), write these values to a file, change θ, and repeat the process. There are also a few other details of taking and storing data, such as keeping up with the increment in θ, the starting and ending values, and the file name in which the data is stored and any heading information which is desired at the beginning of the file. Some extra features may be added which are not absolutely essential but which may save time, energy, and frustration. The most obvious is a simple check to see that the x-ray beam is on, usually by making sure I_o is above some preset minimum. It may be necessary to subtract an offset from I_o and I_t to remove dark counts in a detector or to get above the nonlinear region of a V/F converter. Plotting the data as it is being taken, while it involves somewhat more investment in programming, reveals a wealth of information about any problems (or, hopefully, the lack thereof) in the experiment. To reduce the susceptability to unforeseen events and as a consistency check, several faster scans are better than one very long scan to achieve the same statistical signal to noise ratio. For the experiment to run unattended for long periods, the capability of automatically running several data scans is useful. This will require the program to generate new file names in which to store the additional scans. Since the EXAFS is periodic in electron momentum, a fairly coarse point spacing in θ can be tolerated far above the absorption edge. The ability to vary $\Delta\theta$ over different regions of the data range can greatly reduce the time spent taking data. If the operator discovers some problem with the data it may be desirable to be able to abort the scan early in some reasonably clean manner (other than shutting down and restarting the computer). Provision can also be made to introduce a pause in data collection, although this should be used with caution as it can easily introduce a glitch into the data. Of course, some allowance must be made for the future addition of similar features. Not only will future users have additional ideas, but the computer will inevitably be called upon to manage additional parts of experiments, such as temperature control or changing samples. Such additions could include almost anything, depending on the experiment and the ambitions of the user.

Data manipulation is here taken to be only a few fairly straightforward steps: plotting the data on a CRT terminal or X-Y recorder and transfer to a larger computer for more sophisticated analysis. Alternately, some analysis can be done on the same computer. This is becoming more feasible as both computers and analysis programs become more advanced. The more sophisticated the plotting and analysis capability available on the same computer, the sooner the data can be checked for the characteristics desired in a particular experiment and the sooner any problems can be spotted and corrected. This must be weighed against the complexity of EXAFS analysis and the limits on computing power and time.

Since a computer controlled experiment should be able to run unattended for long periods of time, all parameters necessary for the run must be specified in advance. It may be convenient to save some of these parameters in a disk file so that they need not be continually re-entered by the operator. The parameters associated with the scan are the starting and ending points, the step size and any gradations to be made in the step size, and the integration time or total counts desired for each point. These parameters may be specified in either of two ways: in energy or directly in the monochromator position units (usually stepping motor steps). Specification in energy is easier and is independent of hardware changes, but it is more dependent on conversion accuracies and on the information concerning the hardware being correctly initialized. Specification in monochromator steps is more direct, but must be changed whenever the hardware is changed (the most common being a change of crystal). Whichever method is chosen, the data file should always contain the monochromator positions as a consistency check. Other parameters to be specified are the minimum I_0 to continue taking data (if implemented), any offsets to be subtracted, a file name for the data file, and the number of scans desired (if multiple scanning is implemented). Provision must also be made to input any parameters required by user routines associated with the experiment.

Hardware initialization falls into two categories: initialization of the hardware itself and specification of the constants required by the program to describe the hardware configuration during the running of the experiment. The monochromator calibration includes both types of initialization. The counter for the θ position must be preset to some initial value when the monochromator is at a known position, and this initial value must be communicated to the program. Other constants are also required for conversion of monochromator position to energy, including the lattice spacing of the monochromator crystal.

The parameters setup in advance thus fall into three general categories: 1) hardware parameters which are seldom changed and which correlate with physical changes in the apparatus; 2) scan parameters, which change for different edges and a variety of which should be conveniently stored and ready; and 3) time dependent parameters, which change fairly frequently but are not necessarily correlated with edge changes, for example, offsets, gains, plot scale, etc. These may all be stored in one file or broken up into separate files, depending on the amount of disk space available and the complexity of the program and the experiment.

The convenience routines include any functions necessary for manual control, setting up the apparatus for various experiments, testing it, and solving any problems which arise. It is tempting to scrimp in this area when writing a program or setting up a system, but the price of such economy is often high at some later time. An apparatus which performs beautifully when running but which cannot be repaired without great difficulty will have very limited usefulness. Careful thought should be given to all phases of running the apparatus and sufficient flexibility built into the program to allow for any foreseeable contingency. Some programmers may also wish to include in their programs some functions traditionally left to the operating system, such as listing the disk contents and re-naming files. Such functions may require special, sophisticated techniques and will certainly be system dependent.

HUMAN ENGINEERING

The recent explosion in computing equipment has brought about a radical change in the design emphasis for computer controlled systems such as an EXAFS experiment. The design is no longer limited by the availability of appropriate hardware, as almost unlimited variety and power are available at fairly modest cost. The biggest limit is often finding out precisely what is available and which claims can be believed. Unfortunately, software lags somewhat behind hardware in the explosion. Obtaining or writing quality software thus becomes the major limitation in such a system, and is the area deserving of the most emphasis. Software can now go beyond the most basic functions necessary for the experiment and can be designed for the convenience of the operators and future programmers, as well as some degree of foolproofness. The advantages of such an approach are overwhelming, and may be critical as experiments become more complex. The purpose of this section of the paper is twofold: to act as a starting point for ideas on how to design programs from a user standpoint, and to illustrate some of these design points in hopes that their advantages will become apparent.

For purposes of this discussion, users will be divided into two categories: routine users and systems users. The former are those people who use the apparatus to perform experiments. Their primary concern is that the apparatus do what they want it to with a minimum of input. The system users are those who become more involved with the system, either to maintain it or to modify it for some purpose. Proper design for the system user will be internal, and thus invisible to the routine user.

The routine user will interact with the program's sequence of inputs for taking a data scan. The display seen by the user should be arranged such that all appropriate information is readily available when user input or decisions are necessary. The form of such information and of the input required should be such that an average person will make the right response most of the time, regardless of background, training, lack of alertness, or other factors. This is really a capsule summary of human engineering, and may be somewhat difficult to achieve. An alternate version of the criteria is to avoid unnatural or tricky responses. If the inputs are arranged such that almost any response produces usable results and no possible response has inconvenient or disasterous consequences, a major step toward smooth operation will have been achieved.

Design criteria for the convenience of the system user are not as easily described, but are at least as important since all users are affected by problems encountered or generated by the system user. Some helpful practices are obvious: choice of a simple language, abundant comments, and a carefully planned organization. Modular construction avoids many problems, particularly if it is thoroughly applied. For example, if a move routine is appropriately designed and then always used whenever the monochromator position is changed, then later modifications in the monochromator motion, hardware and/or software, will have fewer repercussions in the remainder of the program. The most difficult part of modularization is deciding where to place the boundaries and how to communicate across them. A final design approach is to make the parameters included in the program to describe the hardware externally changeable. If this approach is carried to its extreme, the routines themselves would contain only procedures; all numbers, option flags, etc. (especially those which are hardware dependent) would be externally alterable. Such external parameters could then be stored in a disk file, for instance. Then any physical changes in the apparatus would require a minimum of changes in the program, perhaps even no changes. Such a complete application of this idea may be restricted by limitations in the programming language.

SUMMARY

Several criteria for the design of a computer - controlled EXAFS experiment have been elucidated, based on experience gained in the construction of such a facility at the University of Washington. Tasks appropriate for physical hardware are fast timing, on the millisecond level and faster, and the retention of critical, volatile information. All other tasks can usually be handled more conveniently in software. Feedback is important on control functions such as the monochromator position. The hardware recommended by the author is a CAMAC system with an LSI-11 microcomputer and eight inch floppy disks. The software should be written as much as possible in a high level language, preferably FORTRAN. The basic software for the experiment is developed, and examples are available from several existing installations; each new designer should be sure to include enough flexibility in auxillary routines to set up and fix their particular apparatus. Finally, and perhaps most importantly, the time has come to devote attention to the human engineering aspects of system design. With the basic experimental software laid out, and with the wealth of hardware available, designers are free to consider the convenience of the users and its role in attaining a smoothly working system.

INSTRUMENTAL ASPECTS OF EXELFS ANALYSIS IN THE ELECTRON MICROSCOPE

D. E. Johnson
Center for Bioengineering, University of Washington
Seattle, WA 98195

S. Csillag
Department of Physics, University of Stockholm
Stockholm, Sweden 11346

E. A. Stern
Department of Physics, University of Washington
Seattle, WA 98195

ABSTRACT

Extended Energy Loss Fine Structure (EXELFS) of high energy electrons can provide the same information as EXAFS and furthermore can provide such information regarding spatially resolved low Z atoms in the laboratory setting using an electron microscope. As an example, a particular experimental configuration is described and preliminary experimental results from two model compounds are shown. The overall characteristics of EXELFS analysis in the electron microscope, including advantages and limitations, are summaried.

INTRODUCTION

Theoretical calculations and recent experimental results[1-5] show clearly that information analogous to that obtained from extended x-ray absorption fine structure (EXAFS)[6] studies can also be found in the energy loss spectrum of fast electrons passing through thin specimins (i.e., Extended Energy Loss Fine Structure ≡ EXELFS). Since, in an electron microspope, this analytical capability can be combined with the spatial resolving power of the electron optical components, the potential exists for spatially resolved EXELFS analysis. Also, laboratory EXAFS instruments as presently contemplated[2] are not able to probe the surroundings of low Z atoms, many of which are technologically important. EXELFS opens the possibility of using EXAFS phenomena to probe low Z atoms in a laboratory facility.

LIMITS OF DISCUSSION

We will discuss only instrumentation for EXELFS analysis in the electron microscope. It is, of course, possible to obtain EXELFS information in an electron beam instrument without imaging capability, but the loss of the spatial characterization and localization features would seem to make this a less desirable alternative.

Regarding the partucular mode of image formation, we will only mention that in the scanning electron microscope mode, smaller beam diameters, and thus higher spatial resolutions are possible than in the conventional transmission electron microscope mode. However, many

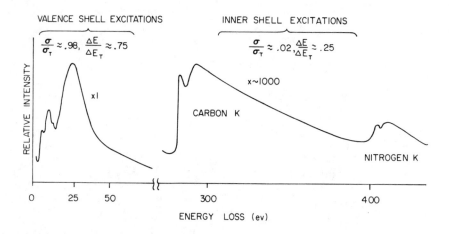

Figure 1 A typical energy loss spectrum of a biological material, in this case, a thin (~400Å) sublimed film of cytosine. Indicated are the fractions of the total cross section (σ/σ_T) and total energy loss $\Delta E/\Delta E_T$ found in the regions of valence shell and inner shell excitations.

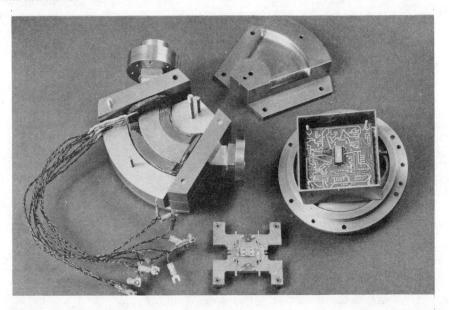

Figure 2 The spectrometer with the top cover plate and pole pieces removed. Shown are; the beam tube, the pole pieces with edges tapered at 45° and a parallel detection system utilizing a photo-diode operated in the electron detection mode.

analytical electron microscopes combine both modes of image formation and either can be used.

A more important distinction between types of microscopes relates to the electron source used. Electron microscopes using a field emission source are characterized by a small energy spread (∼0.2 eV) of the source, and ultrahigh vacuum (∼10^{-10} torr) and by a relatively high cost (∼$700,000). Microscopes using a hot filament source are characterized by a larger energy spread of the source (∼2.0 eV), a high vacuum (∼10^{-6} torr) and lower cost (∼$300,000). These cost estimates are for high spatial resolution (∼5Å) instruments and can be much less if only modest spatial resolution (∼100Å) is required. The typical system and results discussed here, all pertain to a hot filament source system.

GENERAL FEATURES OF THE ENERGY LOSS SPECTRUM

A monoenergetic incoming electron beam has three main regions in the energy distribution after passage through the sample:
a) The unscattered electrons which pass through the sample without losing energy. This is the so-called primary beam or zero loss peak.
b) The inelastically scattered electrons which produce valence shell excitations such as surface or bulk plasmons and lose the corresponding amount of energy necessary to produce this type of excitation. The energy range for these types of excitations is typically between 7-30 eV; and c) The inelastically scattered electrons which produce inner shell excitations losing energies necessary to produce K, L and M shell ionizations.

To illustrate these features, a typical spectrum is shown in Figure 1. The fractions of total cross section (σ/σ_T) and total energy loss ($\Delta E/\Delta E_T$) indicated in each region are only approximate due to the difficulty in extrapolating measured spectra to higher energies. They do serve, however, to indicate the predominance of low energy loss events but with significant energy deposition found in the higher energy loss region.

DESCRIPTION OF A TYPICAL SYSTEM

In order to illustrate the type of instrumentation involved, we describe briefly here the system we have been using in a feasibility study of electron microscope based EXELFS analysis.

The experimental apparatus used consists of a JEOL 100C transmission electron microscope with scanning attachment and magnetic sector electron energy loss spectrometer, interfaced to a Kevex 7000 computer based multichannel analyzer for data gathering and initial processing. The design and construction of the energy loss spectrometer and its interfacing to the electron microscope and multichannel analyzer are described in detail elsewhere [7] and only a brief summary of the characteristics are given here.

The spectrometer is a straight edge magnetic sector (double focusing) spectrometer with a radius of curvature = 5 cm and uniform magnetic field = 224 gauss which produce a dispersion = 1μm/eV at 100 KeV, bending angle = 90°, object and image distances = 10 cm, magnification = 1. See Figures 2 and 4.

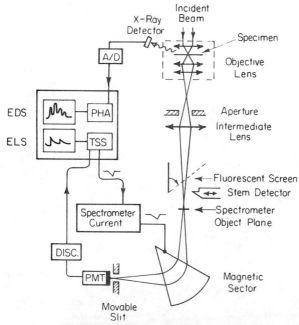

Figure 3 A schematic diagram of the energy loss microanalysis system. Electrons of a given energy loss are counted and stored in a multichannel analyzer operated in the time sequence store mode (TSS). In this mode, the analyzer provides an analog ramp, synchronized with the sweep through the memory channels, which is used to ramp the magnet current producing an energy loss spectrum.

Figure 4 The spectrometer assembly mounted in the knee well of the electron microscope. Above the spectrometer is the retractable scintillator-light pipe-photomultiplier detector for STEM.

Electrons passing through an energy selecting slit strike a plastic scintillator (pilot B) and the resulting photons pass down a lucite light pipe to a photomultiplier tube (RCA 8895). The anode pulses of this tube pass directly to a discriminator (Ortec Model 436, 100 MHz), the output pulses of which are counted and stored in a multichannel analyzer (Kevex 7000) operated in the time sequence store mode. An analog ramp produced by the analyzer in this mode is used to ramp the magnet current supply (Kepco-JMX 6-5M) thus sweeping the energy loss spectrum across the energy selecting slit synchronously with the analyzer's sweep through the memory channels.

A schematic diagram of the instrument is shown in Figure 3. Also indicated in this figure is an energy dispersive X-ray analysis system which the energy loss system is intended to complement in elemental microanalysis studies.

As shown in Figure 3, the objective lens field beneath the specimen and the intermediate lens are used to produce an image at the spectrometer object point of the specimen area illuminated. This lens combination can operate over a range of magnifications for these same two conjugate points. A detailed analysis of the electron optics of this combination [8] has shown that for a given diameter of the specimen area selected and a given angular acceptance into the system, there exists optimum operating modes of the lens system which result in the minimum spectrometer image size and thus the best energy resolution.

The key area of instrumental development remaining is that of a parallel data collection system. Our initial experiments with a photodiode array operated in the electron detection mode indicated that, for our requirements, photon detection would be more desirable (thus eliminating, for example, radiation damage of the array). The system presently being assembled for testing uses the same photodiode array but now coupled to a transmission phosphor conversion plate (electrons to photons) via lenses and a small image intensifier (gain $\sim 10^4$). The system will be installed on the microscope in addition to the variable width slit detection system now in use. For those applications where the highest sensitivity is required, the slit will be opened and a magnified ($\sim 20X$) region of the energy loss spectrum will be imaged on the conversion plate of the parallel system by a magnetic lens. This magnification is necessary to match the dispersion of the spectrometer ($\sim 1\mu m/eV$) with the resolution of the diode array ($\sim 25\mu m$).

The increased sensitivity of this parallel detection system ($\sim 200X$) will reduce the effects of radiation damage and contamination allowing increased spatial resolution.

TYPICAL EXELFS ANALYSIS

In order to demonstrate the feasibility of the technique, we have studied a number of different systems with different ionization edges. We illustrate here with results from thin films of Al_2O_3 and MgO.

The Al_2O_3 samples were prepared as thin Al evaporated on the surface of a freshly cleaved NaCl crystal at 20°C. The films were

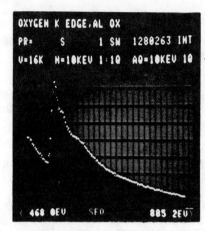

Figure 5 Typical EXELFS past the O K-edge in the energy loss spectrum of a thin Al_2O_3 film. The incident energy was 100 KeV, the half angle of acceptance was 9mr, and the data gathering time was 200 seconds.

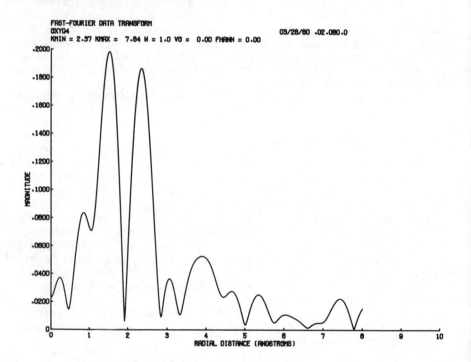

Figure 6 The magnitude of the Fourier transform of $k\chi(k)$ data from the O K-edge in the energy loss spectrum of Al_2O_3 of Figure 5. Correction for the phase shift is not included.

oxidized by heating in air. Al_2O_3, being an insulator, charges up during exposure to the electron beam, and, in order to avoid this, we evaporated a thin (~30Å) carbon film onto the Al_2O_3. Floating the film off the NaCl in water, we obtained thin, carbon-coated Al_2O_3 films with an estimated thickness of about 200Å. The MgO specimens were prepared in a similar fashion. Electron diffraction patterns obtained from these specimens showed diffuse amorphous patterns from the Al_2O_3 films and polycrystalline ring patterns from the MgO films. Electron micrographs, electron diffraction patterns and electron energy loss spectra were all obtained in the same instrument from the same part of the sample. Figures 5 and 7 show typical EXELFS past the O K-edge in Al_2O_3 and MgO. The data were obtained with an accumulation time of 1 sec/data point for a total time of approximately 200 sec. The incident beam energy = 100 KeV, the beam current was 4×10^{-9}A and the acceptance half angle = 9 mr. The instrumental energy resolution for the above spectra was approximately 5 eV determined mainly by lens aberrations and the size of the irradiated area [8]. With the more efficient parallel detection system, the beam current and thus the size of the irradiated area could be reduced resulting in an energy resolution of approximately 2-3 eV. By using alternative electron sources (e.g. field emission tungsten) the resolution could be further reduced to <1 eV.

DATA ANALYSIS AND RESULTS

The data analysis involves several steps and is very similar to the familiar EXAFS data analysis, namely: smooth background subtraction, normalization of the modulations to the edge step height and converting energy loss to ejected electron wave vector. For the Al_2O_3, the useful data contraction was started at 30 eV past the edge and 40 eV past the edge for the MgO data, to minimize any effect of plural scattering.

Figures 6 and 8 show typical Fourier transforms of $k\chi(k)$ data from the oxygen K edges of Figures 5 and 7. The zero of the ejected electrons wave vector was taken at the inflection point of the edge.

GENERAL FEATURES OF EXELFS ANALYSIS IN THE ELECTRON MICROSCOPE

1) Advantages of EXELFS analysis: a) EXELFS permits the extraction of fine structure information from low Z elements, which is a difficult problem for synchrotron radiation EXAFS; b) One can focus the electron beam to very small areas (in principle as small as ~5Å in diameter) providing spatial resolution to study inhomogeneous samples; c) Using the electron microscope one can image the irradiated area and also obtain the diffraction pattern of the sample, both of which contain useful additional information; d) The data gathering time is certainly comparable to that of synchrotron sources; e) One can study the momentum transfer dependence of the inelastic electron scattering cross section; f) Finally, the instrumentation is more accessible and less expensive than synchrotron sources.

2) Limitations of EXELFS analysis: a) In the X-ray EXAFS the sample thickness affects only the amplitude of the modulations (typically 3-5% of the edge step height) but in the EXELFS case the

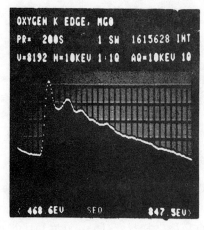

Figure 7 Typical EXELFS past the O K edge in the energy loss spectrum of a thin MgO film. The incident energy was 100KeV, the half angle of acceptance was 9mr and the data gathering time was 200 sec. The spectrum was smoothed once.

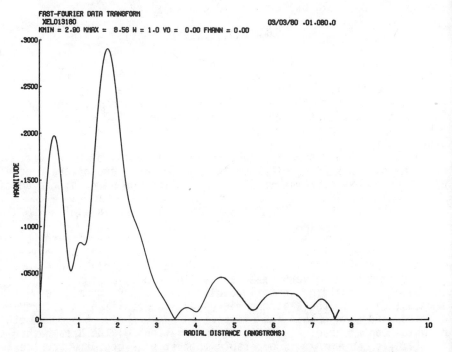

Figure 8 The magnitude of the Fourier transform of k $\chi(x)$ data from the O K edge in the energy loss spectrum of MgO of Figure 7. Correction for the phase shift is not included.

thickness affects both the amplitude and can also add multiple scattering peaks to the modulations by the convolution of the low lying energy loss spectra with the core excitation spectra. There is an optimal sample thickness depending on the energy of the bombarding electrons and this optimal thickness is approximately equal to that required for high resolution transmission electron microscopy; b) The maximum specimen dose (electrons/cm^2) may be limited for sensitive specimens by radiation damage; c) The maximum specimen dose may be limited by specimen contamination if other than ultra-high vacuum systems are used; d) The maximum data range may be limited by K ionization edges which are too close to each other such as carbon and oxygen K-edges; e) If the energy loss measurements are made at high angular resolution, for example in the study of anisotropic specimens, signal levels will decrease significantly over those found in this work using large angles of collection.

Limitation (a) above can be eliminated by the use of sufficiently thin specimens as have been used in this work. The effects of limitations (b),(c) and (e) above can all be reduced by the development and application of parallel data collection systems (i.e., all energy loss intervals detected simultaneously). Such a system could have increased the counting rates in the work reported here by a factor of $\simeq 200$. Such an increase in sensitivity can then be used to reduce the specimen dose, to increase the angular resolution, or to study time dependent processes with EXELFS.

SUMMARY

The ability to study extended fine structure from low Z materials with the spatial resolution and imaging possibilities provided by the electron microscope is unique, and would appear to make this technique a powerful tool in the study of low Z compounds such as biological samples.

ACKNOWLEDGEMENTS

This work was supported by NIH grants: HL21371 and HL00472 (RCDA, D. Johnson) by a Swedish Government scholarship (S. Csillag) and NSF grant PCM 79-03674 (E. A. Stern).

REFERENCES

1. Kincaid, B. M., Phys. Rev. Lett 40, 19 (1978).
2. Batson, P. E. and A. J. Craven, Phys. Rev. Lett 42, 892 (1979).
3. Leapman, R. D. and V. E. Cosslett, J. Phys. D. Appl. Phys., 9, (1976).
4. Johnson, D. E., Csillag, S., Stern, E. A., Proc. EMSA, 37, 526 (1979).
5. Isaacson, M. and M. Utlaut, Optic 50, 213 (1978).
6. Stern, E. A., Phys. Rev. B 10, 3027 (1976).
7. Johnson, D. E, Rev. Sci. Inst. 51, 705 (1980).
8. Johnson, D. E., Ultramicroscopy, 5, 163 (1980).

EDITOR'S NOTES - CHAPTER 6

z. The x-ray laboratory EXAFS facilities as described in the rest of the chapters of this Proceedings are capable of measuring EXAFS down to x-ray energies of 2-3 KeV. X-ray spectrometers capable of covering lower energy ranges have been built and at least one has been employed to measure EXAFS in that range.* To cover such a soft x-ray range requires enclosing the x-ray source, crystal monochromator, sample and detector all in vacuum so as to eliminate as much mass as possible from absorbing the soft x-rays. The intensity I_0 for soft x-ray facilities is much less than the ones described here because of difficulties in the generation of soft x-rays without harmonics as described in Chapters 2 and 8 and in the low efficiency of reflection of crystals with the large d spacing required for monochromating the soft x-rays.

*S. Kiyono, Y. Hayasi and T. Muranaka: <u>International Conference on the Physics of X-Ray Spectra</u>, (Gaithersburg, Maryland, 1976); S. Kiyono, S. Chiba, Y. Hayasi, S. Kato and S. Mochimaru, Proc. International Conf. on X-ray and XUV Spectroscopy, Sendai, 1978; Japanese Journal of Applied Physics, <u>17</u>, Supplement 17-2, pp. 212-214 (1978).

COMPARISON OF LABORATORY AND SYNCHROTRON RADIATION EXAFS FACILITIES

Ruprecht Haensel
Institut für Experimentalphysik der Universität Kiel,
Olshausenstr. 40-60, D-2300 Kiel 1, F.R. Germany

ABSTRACT

Although synchrotron radiation with its unique properties as an X-ray radiation source gave a decisive impetus to the development of EXAFS (extended X-ray absorption fine structure) spectroscopy the increasing number of synchrotron radiation facilities is matched by a similar trend for laboratory EXAFS facilities. The mutually complementary roles of both types of facilities will be reviewed.

INTRODUCTION

Figure 1 shows an EXAFS spectrum of metallic Cu in the vicinity of the K-absorption edge. Depending on the radiation source used for the experiment it can be recorded in hours, minutes or nanoseconds. The first time interval is related to the use of an in-house facility with a (fixed or rotating anode) classical X-ray generator and a flat crystal monochromator, the second to the use of synchrotron radiation and the last to a laser-generated plasma used as an X-ray source. The choice of the best source and facility for a given experiment depends on several details encountered in the problem to be investigated. Therefore we have to discuss some essentials of EXAFS spectroscopy before we review the different facilities in the next section. For a more general introduction into the basic principles

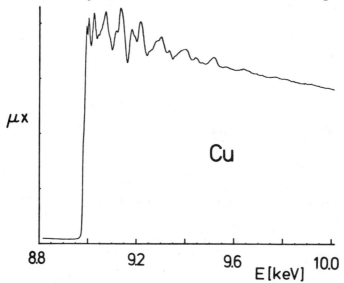

Fig.1 Absorption coefficient of polycrystalline Cu at the K-edge[1].

and the different fields of applications of EXAFS spectroscopy the reader is referred to several reviews in the literature [2-8].

Although the first observations of EXAFS spectra and the qualitative explanation of its origin by Kronig[9] have been made fifty years ago, it is only less than a decade since Stern, Lytle and Sayers[10] developed a method to extract quantitative geometrical structure informations from the experimental data. According to their and other authors' theoretical work [10-12] the EXAFS modulation $\chi(k)$ can be written as

$$\chi(k) = \sum_j A_j(k) \cdot \sin(2kR_j + \Phi_j(k)) \tag{1a}$$

with the amplitude function

$$A_j(k) = -\frac{1}{k} \cdot \frac{N}{R_j^2} |f_j(\pi,k)| \cdot \exp(-R_j/\lambda) \cdot \exp(2\sigma^2 k^2) \tag{1b}$$

and the phase term

$$\Phi_j(k) = 2\delta_1(k) + \arg f_j(\pi,k) \tag{1c}$$

The summation in Eq. (1a) is carried out over all coordination shells j; N_j is the number of identical atoms in the scattering shell j at the distance R_j; $|f_j(\pi,k)|$ is the backscattering amplitude characteristic for the scattering atom species; $\Phi_j(k)$ is the scattering phase consisting of the phase shift $\arg f_j(\pi,k)$ due to the potential of the scattering atom and the phase shift $2\delta_1(k)$ due to the potential of the central atom. The first exponential in eq. (1b) describes the finite range of the photoelectron wave, typically 3 to 6 coordination shells in extension and the second exponential is a Debye-Waller factor, which takes into account the dynamic (thermal) or static (structural) disorder with the mean square relative displacement σ^2. It is mainly this term which is responsible for the damping of the EXAFS modulation at high k-values, but for light scatterers also the backscattering amplitudes vanish for high k-values.

Since EXAFS stems from a diffraction of a photoelectron wave with the excited atom serving as an internal electron source it is appropriate to write χ as a function of the wavenumber k, with

$$k = \sqrt{\frac{2m}{\hbar^2}(E-E_o)} \tag{2}$$

where the difference between the exciting photon energy E and the binding energy E_o of the electron with mass m is the kinetic energy of the photoelectron.

Experimentally χ is extracted from the characteristic absorption μ above the edge according to

$$\mu_c = \mu_o(1+\chi) \tag{3}$$

where μ_o describes the smooth absorption of the free atom. μ_c however has first to be recovered from the total absorption μ by subtraction the background absorption

$$\mu_c = \mu - \mu_b \tag{4}$$

In monoatomic samples μ_b stems from the outer shell transitions, in multiatomic samples transitions from the other atom species add to μ_b.

From experimental $X(k)$ data the structural information can be gained in several ways: The simplest case is when only one scattering shell contributes to $X(k)$. In this case parametrized envelope functions $A(k)$ and phases $\Phi(k)$ can be fitted to the experimental data. The increasing number of parameters in systems with more than one scattering shell limits the applicability of these techniques, although in favourite cases the next neighbour's distance can still be taken immediately from the EXAFS spectrum, as Mallozzi et al.[13] have shown for metallic Al. In general however a Fourier transform of $X(k)$ is neccessary to obtain the radial distribution function $F(r)$ in real space.

The quality of the $F(r)$ data depend on the quality of the $X(k)$ input. Generally the statistical accuracy of absorption measurements ($I = I_0 \cdot \exp(-\mu x)$) depends on the accuracy of the measurement of transmitted light intensity I [14,15]. The optimum thickness of the sample is $x_c = 2/\mu$ (corresponding to a transmission of ~10%), which leads to typical values of $x_c = 1$ to 5 μm for absorbers having their characteristic absorption in the 10 keV-regime. Consequently the statistical uncertainty of μ_c and hence for X amounts to

$$\frac{\Delta X}{X} = \left(\frac{e}{2}\right) \cdot \left(\frac{\mu}{X}\right) \cdot \left(\frac{1}{I_0}\right)^{1/2} \tag{5}$$

As we see from Eq.(5) there are two ways to optimize the data: i) For a given intensity X can be improved by cooling (Debye-Waller term in Eq.(1b)) and by keeping μ_b low; ii) For given absorption conditions the incident intensity I_0 should be as high as possible to keep the data collection time within reasonable limits. The statistical uncertainty determines the upper k-limit covered by the measurement (the lower k-range limit is set at k 2 Å^{-1} because of the overlap with the density of states structures near the edge) and hence the spatial resolution of $F(r)$.

In a straightforward absorption experiment the I_0 and I are measured with and without the sample in the beam consecutively or simultaneously. The absorption of the X-ray photons in the sample is followed by fluorescence radiation from the excited atom and/or by emission of photoelectrons. The fluorescence light and photoelectron energies and yields are characteristically dependent on the excited atom species and for a given system they closely follow $X(E)$. Therefore X can be gained from these secondary products as well and this has several advantages: Inhomogeneities of the sample thickness and higher harmonics transmitted from the monochromator do not affect the fluorescence and photoelectron yield. The presence of other atoms in highly dilute samples contributes substantially to μ_b in absorption measurements (up to the complete disappearance of the characteristic absorption) but does not contribute to the characteristic fluorescence signal. Photoelectrons containing the $X(k)$ information originate from surface regions, therefore investigation of surface layers (adsorbates) are possible, discriminating against the background signal from the substrate. The photoelectrons can be produced in different ways: One is the direct production during the photoabsorption process. Another is an Auger decay of the excited atom. Alternatively the excited atom may also recombine under emission of a fluorescence photon which leads to photoemission from another atom. The different channels result in different escape depths of the photoelectrons of up to 1ooo Å and more [16].

The application of these secondary EXAFS methods to dilute samples allows substantial improvements of the ratio of the characteristic signal to unwanted background,[x] but since the efficiency for the production of secondaries is rather small, a high I_o is required to obtain a sufficient statistical accuracy. These are the cases where synchrotron radiation extends its full potential in allowing to study absorbates in sub-monolayer thicknesses and dilute systems in sub-ppm concentrations.

Whereas the study of adsorbates and surface regions by photoelectron yield EXAFS is only possible with synchrotron radiation, a new method has been developed by Martens [17] to study superficial regions by reflectivity measurements with a laboratory EXAFS facility. The reflectivity of a sample is determined by the complex index of refraction $n = 1 - \delta - i\beta$. Below the critical angle of total reflection the energy dependence of the reflectivity is mainly ruled by the immaginary part β, i.e. the absorption coefficient [18]. Since the penetration depth of the light is only 2o to 5o Å, we can obtain EXAFS signals from surface regions. The penetration depth can be varied by changing the angle of incidence.

RADIATION SOURCES

The first EXAFS measurements by Lytle et al. [19] were performed with normal X-ray generators. In our group EXAFS spectroscopy started 1975 with a 12 kW Rigaku source [2o]. For such a source equipped with a flat crystal monochromator and a NaI scintillation counter typical scanning times for an absorption spectrum are ~1o hours. At about the same time the first experiments using synchrotron radiation were started at the SSRL in Stanford, soon followed by those at LURE in Orsay, NINA in Daresbury and DESY in Hamburg [21]. The new sources allowed scanning times to be cut down to minutes. At the beginning this short scanning time was not used to extend the scientific program to short-living samples, but to increase the throughput of large series of samples. The main progress however was, that the use of synchrotron radiation allowed the development of the secondary EXAFS methods mentioned in the last section.

Synchrotron radiation did not supersede the use of in-house facilities. On the contrary: the interest in EXAFS spectroscopy grew so rapidly and improvements of in-house facilities, bringing scanning times down to fractions of an hour, were so substantial, that absorption measurements and fluorescence measurements on moderately dilute samples can now be made at home with sufficient quality. On the other hand, in-house facilities are ideally suitable to prepare and optimize equipment and samples before they go to a synchrotron radiation center. In this way the most economic use of accelerators can be made and the scarce machine time can be reserved for experiments which absolutely need it.

A comparison of synchrotron radiation and X-ray generators shows that X-ray generators emit isotropically whereas synchrotron radiation is confined in a small aperture of typically 0.1 mrad at 10 keV photon energy. The spectral distribution of synchrotron radiation is completely flat. On the other hand X-ray generators, besides the bremsstrahlung, produce characteristic emission lines, originating not only from the anode material, but also from impurities (mainly W from the heater) which grow during long time operation of the source. These lines can cause troubles in the data reduction. Knapp[22] uses a feedback from a reference detector to the current control of the source resulting in a complete flattening out of emission lines. This procedure can not so easily be performed with the Rigaku source, but sufficiently accurate reproducibility of the monochromator positioning and linearity of the detectors also allow complete cancellation of disturbing influences of the emission lines.

A quantitative comparison of the bremsstrahlung of an X-ray generator (Cu anode, 45 keV, 500 mA) and the synchrotron radiation of DESY is shown in Fig.2 [23]. For a given aperture at the synchrotron (distance source-monochromator 37 m, entrance window height 5 mm ~ 0.15 mrad and width 40 mm ~ 1 mrad) the brightness of synchrotron radiation clearly surmounts that of the bremsstrahlung by many orders of magnitude, if we assume the same angular aperture (0.15 mrad^2) for the X-ray generator. For a laboratory arrangement using a flat

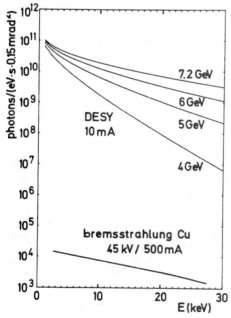

Fig.2 Comparison of the intensities of Synchrotron radiation at DESY and the bremsstrahlung of a Cu-anode [23].

crystal monochromator however the aperture can easily be made larger by a factor of 100 and for focussing crystal monochromators by another factor of 100 to 1000. On the other hand the numbers given in Fig.2 are for the synchrotron DESY. At storage rings the intensity is higher by at least one order of magnitude because of the higher current in the machine and can be even further increased by a insertion of wigglers and undulators.

The decisive quantity in any case is the number of photons available at the experiment. Using conventional sources and flat crystal monochromators count rates in the order of 10^4 cps can be obtained [20], with focussing crystal monochromators they can be increased to 10^6 to 10^7 cps [24,25] and for storage rings values of 10^{12} cps have been reported [26].

Synchrotron radiation is linearily polarized in the electron orbital plane. Above and below this plane additional vertical components come up, which add to an overall elliptical polarization of the radiation. The degree of polarization depends on the angular range of radiation used in the experiment. It amounts to more than 80% for typical exit slits of several mm height. The polarization has been used for EXAFS studies of anisotropic materials. For X-ray tubes on the other side the degree of polarization is difficult to calculate. It depends on the wavelength, the take-off angle from the anode, the anode material and the acceleration potential.

X-ray generators normally have a high constancy in time. Electron accelerators on the other hand are pulsed light sources with pulse lengths of 0.1 to 1 nsec and a pulse separation of 2 nsec to 1 μsec depending on the mode of operation. Different proposals exist to increase the pulse separation further with the help of electron beam steering devices and with mechanical chopper wheels. This time structure has to be taken into account in view of the time resolution and linearity of the detectors. On the other hand it has very promising capabilities for future experiments. We have already mentioned the laser-generated plasma source, which has been demonstrated by Mallozzi et al.[13] to make EXAFS absorption measurements with one single shot of 2 nsec duration.[v] Unfortunately the repetition rate for this source is rather low. Synchrotron radiation on the other hand permits the accumulation of time-resolved data over many cycles in a very short time. This has been demonstrated by Huxley et al.[27], who obtained electronically recorded time-resolved diffraction patterns of a frog muscle. Time-resolved spectroscopic investigations of luminescence emission also became a standard technique with synchrotron radiation [28]. With these measurements excited electronic states are probed, but some theories to explain the nature of the emitting centers include localized dynamic lattice distortion and relaxation effects, which could be a case for EXAFS in the foreseeable future. First steps have been done at Stanford, where EXAFS spectra were obtained with exposure times of less than 1 sec [29].

MONOCHROMATORS AND DETECTORS[u]

Here we find no basic differences between in-house and synchrotron radiation facilities. Variations in the equipment are rather a function of the financial possibilities.

The simplest monochromator arrangement consists of a flat crystal in the beam, mounted on a goniometer with the θ-drive for the crystal and the 2θ-drive for the detector (Fig.3a). The geometrical energy resolution Δ E of this arrangement is

$$\Delta E = E \cdot \Delta \theta \cdot \cot \theta \qquad (6)$$

with the angular resolution

$$\Delta \theta = \frac{s+d}{a} \qquad (7)$$

where s is the (projected) source width, d the detector width and a the distance between the source and the detector. A practical case may be s = d = 50 µ , a = 0.5 m (for an in-house facility), θ = 20° at E=9000 eV (near the Cu K-edge) resulting in ΔE ~ 5 eV. This is sufficient resolution for EXAFS-spectroscopy, the instrinsic crystal resolution does not affect the result.

If the detector arm can not be easily rotated, a double crystal mount (Fig.3b) allows a wavelength scan with a fixed exit beam direction. However the exit beam is displaced relative to the incoming beam by

$$V = 2D \cdot \Delta \cos \theta \qquad (8)$$

with D being the distance of the parallel crystal surfaces. Because

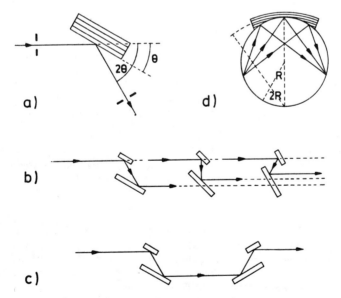

Fig.3 Different monochromators for EXAFS-experiments:
 a) one flat crystal, b) double crystal
 c) 2 double crystals, and d) focussing crystal arrangement.

of this displacement the sample chambers and detectors have to be shifted accordingly. The displacement can be avoided by inserting a second pair of crystals in the beam or by an additional translational movement of the second crystal (Fig.3c).

To gain intensity curved crystal arrangements (Fig.3d), can be used. Experiments have been made with Johansson [24,25] and with Johann [30] mountings. The geometrical energy resolution is the same as given by Eq.(6), but for the Johansson mounting problems can arise because of imperfect crystal grinding and mounting [25]. For both mountings the crystal and the detector move along a circle through the source. The distances between the source and the crystal and the crystal and the detector have to be kept equal but are simultaneously changing while scanning over the spectrum. This can be realized by a rhomboid mounting [24] or by an eccentrically mounted goniometer [25]. In both cases the crystal eventually can leave the emission cone of the source, unless the source can be rotated around its focal axis [31]. Most of the monochromators used so far are working in air. X-ray absorption of air limits the low energy range to about 6 keV. Crystals used so far were LiF (200), Si (111), Si (220), Ge (111), Ge (220) and Ge (311). With evacuated monochromators using Bragg reflection the energy range can be extended down to ~1 keV [32-34]. In the energy range below 1000 eV EXAFS measurements have been performed with a glass grating monochromator [35].

Because of the larger distances in synchrotron radiation facilities the focussing and the dispersion of the radiation are separately done by concave mirror arrangements and flat crystals respectively. A problem not yet completely solved is the heating of the first optical element by the synchrotron radiation and the question whether the first element should be the crystal or the mirror.

The most widely used detectors for X-ray radiation are NaI scintillation counters, Ge(Li) solid state counters and ionization chambers. Scintillation counters were mostly used for low intensity absorption measurement with laboratory facilities. They and solid state counters are also used for fluorescence light detection. Here a discrimination against unwanted photons (elastically scattered photons from the same element, fluorescence from other elements) is important. It can be done through the energy resolution of the solid state counters or for the scintillation counters by pulse height discrimination, and alternatively by a secondary monochromator or by edge-filters between the sample and the detector. For electron detection various types of open photomultipliers are used.

CONCLUSIONS

EXAFS spectroscopy grew rapidly in the last years from an interesting physical phenomenon to a daily working tool in the investigation of the local structure in various systems. In these investigations in-house and synchrotron radiation facilities are playing ideally complementary roles. The enthusiasm about this method however may not obscure our view to the other methods of structure investigations: If we increase the energy resolution of our spectrometers to ~1 eV we automatically record together with our EXAFS spectra the near-edge structures. For the evaluation of the EXAFS data they have to be discarded, but they contain much additional information on the electronic (chemical) state of the central atom which is often directly connected to the geometrical structure. Comparisons between EXAFS, Mössbauer Spectroscopy and Muon Spin Rotation have been recently made by Chappert [36]. Last but not least we should not forget neutron and X-ray diffraction [37]. Between EXAFS and X-ray diffraction there are the closest connections but also substantial differences, mainly in the k-range, where most of the information is carried in both methods. Differences in distances and coordination numbers obtained from both methods allow important conclusions as to the symmetry or asymmetry of radial distribution functions [37]. The possibility to select the central atom in EXAFS spectroscopy finds its correspondence in the anomalous scattering method. Here synchrotron radiation can play an important role by offering the possibility to select the optimal exciting energy without being confined to the characteristic lines of an X-ray generator [38].

ACKNOWLEDGEMENTS

The work of our group is supported by the Bundesministerium für Forschung und Technologie through the Deutsches Elektronen-Synchrotron DESY and by the Deutsche Forschungsgemeinschaft. Thanks are due to my coworkers P. Rabe, W. Böhmer, R. Frahm, G. Martens, D. Mucha, W. Pronkow, G. Tolkiehn, W. Thulke, P. Wenck, and A. Werner for a fruitful collaboration and for many stimulating discussions. Mrs. M. Behrens kindly prepared the manusript.

REFERENCES

1. W. Böhmer and P. Rabe, J.Phys. C12, 2465 (1979).
2. L. V. Azaroff, Rev. Mod. Phys. 35, 1012 (1963).
3. S. P. Cramer and K. O. Hogdson, in: Progr. Inorg. Chem. Vol. 25, ed.: S. J. Lippard (J. Wiley, 1979), p. 1.
4. P. Eisenberger and B. M. Kincaid, Science 200, 1441 (1978).
5. G. S. Knapp and F. Y. Fradin, in: Electron and Positron Spectroscopies in Materials Science and Engineering, eds.: O. Buck, J. K. Tien, and H. L. Marcus, (Adacemic Press, 1979), p. 243.
6. D. R. Sandstrom and F. W. Lytle, Ann. Rev. Phys. Chem. 30, 215 (1979).
7. E. A. Stern, Contemp. Phys. 19, 289 (1978).
8. P. Rabe and R. Haensel, in: Advances in Solid State Physics, Vol. XX, ed. J. Treusch, (Pergamom/Viewes, 1980), p. 43.
9. R. de L. Kronig, Z. Phys. 70, 317 (1931) and 75, 191 + 468 (1932).
10. E. A. Stern, Phys. Rev. B 10, 3027 (1974).
    F. W. Lytle, D. E. Sayers, and E. A. Stern, Phys. Rev. B12, 4825 (1975).
    E. A. Stern, D. E. Sayers, and F. W. Lytle, Phys. Rev. B12, 4836 (1975).
11. C. A. Ashley and S. Doniach, Phys. Rev. B11, 1279 (1975).
12. B.-K. Teo and P. A. Lee, J. Am. Chem. Soc. 101, 2815 (1979).

13. P. J. Mallozzi, R. E. Schwerzel, H. M. Epstein, and B. E. Campbell, Science 206, 353 (1979).
14. L. G. Parratt, C. F. Hempstead, and E. L. Jossem, Phys. Rev. 105, 1228 (1957).
15. J. J. Jaklevic, J. A. Kirby, M. P. Klein, A. S. Robertson, G. S. Brown, and P. Eisenberger, Sol. State Comm. 23, 679 (1977).
16. G. Martens, P. Rabe, N. Schwentner, and A. Werner, J. Phys. C11, 3125 (1978).
    W. Pronkow, Diplomarbeit Univ. Kiel (1980).
17. W. Martens, Dissertation Univ. Kiel (1980).
18. G. Martens and P. Rabe, phys. stat. sol.(a) 58, 415 (1980)
19. F. W. Lytle, in: Developments in Applied Spectroscopy, Vol.2, eds. J. R. Ferraro and J. S. Ziomek (Plenum Press, 1963), p. 285.
    D. E. Sayers, F. W. Lytle, and E. A. Stern, J. Non-Cryst. Sol. 8-10, 401 (1972).
20. G. Martens, P. Rabe, N. Schwentner, and A. Werner, Phys. Rev. Letters 39, 1411 (1977).
21. For more complete informations the reader is referred to the publication lists of the different synchrotron radiation laboratories.
22. G. S. Knapp, see contribution to this volume (Chapter 1).
23. P. Rabe, G. Tolkiehn, and A. Werner, Nucl. Instr. Meth. 171, 329 (1980).
24. P. Georgopoulos and G. S. Knapp, J. Appl. Chryst. (to be published).
25. R. Frahm, R. Haensel, D. Mucha, P. Rabe, and W. Thulke, (to be published).
26. J. B. Hastings, B. M. Kincaid, and P. Eisenberger, Nucl. Instr. Meth. 152, 167 (1978).
27. H. E. Huxley, A. R. Faruqui, and J. Bordas, M. H. J. Koch, and M. Milch, Nature 284, 140 (1980).
28. N. Schwentner, Appl. Opt. (to be published).
29. T. Matsushita, private communication (also Chapter 9).
30. K. Lu, see contribution to this volume (Chapter 9).
31. E. A. Stern, private communication.
32. M. Lemonnier, O. Collet, C. Depautex, J.-M. Esteva, and D. Raoux, Nucl. Instr. Meth. 152, 109 (1978).
33. P. Wenck, Diplomarbeit Univ. Kiel (1980).
34. J. Stöhr, private communication.
35. J. Stöhr, L. Johansson, I. Lindau, and P. Pianetta, Phys. Rev. B20, 664 (1979).
36. J. Chappert, J. de Physique, Coll., 41, C1-9 (1980).
37. C. N. J. Wagner, J. Non-Cryst. Sol. 31, 1 (1978).
    R. Haensel, P. Rabe, G. Tolkiehn, and A. Werner, in: Liquid and Amorphous Metals, eds.: E. Lüscher and H. Coufal,(Sijthoff & Noordhoff, 1980), p.
38. P. H. Fuoss, W. K. Warburton, and A. Bienenstock, J. Non-Cryst. Sol. 35-36, 1233 1980).

EDITOR'S NOTES - CHAPTER 7

z. This value of $\mu x_c = 2$ is valid if no monitoring of I_o is necessary. As pointed out in Chapter 4 when I_o monitoring is necessary, as is usually the case, the optimum $\mu x_c = 2.6$.

y. A more detailed discussion of statistical noise is given in Chapters 1 and 4.

x. See also discussion in Chapters 1 and 4.

w. See also Chapter 1.

v. This material is also presented in Chapter 9.

u. Further details are presented in Chapters 3, 4, 9 and 10.

SUMMARY

B. Ray Stults (Chairman)
Monsanto Co., 800 N. Lindbergh, St. Louis, MO 63166

An individual workshop dealing with x-ray sources for the in-house EXAFS laboratory was held as part of the Workshop on Laboratory EXAFS Facilities and Their Relation to Synchrotron Radiation Sources. The workshop panel members were: John Holbin from Marconi-Elliott Avionics Ltd., David Hempstead from Rigaku/USA Inc., Dr. J. Azoulay from the University of Washington, and Drs. R. E. Schwerzel and P. T. Mallozzi from Battelle, Columbus Labs. The workshop did not attempt to evaluate the various x-ray sources as to which was best for the in-house laboratory but rather attempted to determine the unique properties of each laboratory x-ray source. As was discussed throughout the workshop conference, the x-ray source is only part of the total system and the experimenter must evaluate which type of source is best to complement the anticipated experiments. For example, if one needs to avoid possible higher order harmonics during data collection, then an x-ray generator which operates at low accelerating voltages (less than 20 KeV) with maximum current is very desirable. It is also important, in many instances, to have a choice of anodes in order to avoid potential characteristic lines which may appear in a critical energy region of the EXAFS spectrum. Below is a brief summary of the workshop followed by individual panel member contributions dealing in more detail with selected topics. There is also a contribution from Dr. G. Via from Exxon Company, discussing the use of a scanning electron microscope as an x-ray source.

There were identified during the workshop on sources and the general conference three different classifications of x-ray sources not including synchrotron radiation[1]. The three classifications are: (1) conventional x-ray generators, both sealed tube and rotating anode, (2) focusing sources such as the scanning electron microscope, and (3) soft x-ray sources including cold-cathode x-ray tubes and laser-produced x-rays.

Conventional x-ray generators, both sealed-tube and rotating anode, represent the most common x-ray sources for general in-house EXAFS experimentation. Sealed tube x-ray generators nominally produce from 1.0 to 1.5 kW of power and they represent the most easily accessible and most economical x-ray sources for laboratory EXAFS. They are usually very stable sources of x-radiation and are available with a variety of target materials. In addition, sealed tube generators are generally available in most academic or industrial institutions. Their main disadvantage is the relatively low photon flux as compared to rotating anode sources[2]. Rotating anode x-ray generators were identified as producing the highest photon flux with the widest energy range for laboratory EXAFS. The low energy cut-off is governed by the thickness of the beryllium window and the air absorption of the x-rays. The usable high-energy cut-off is controlled by the resolution of the crystal monochromators but it was generally felt that EXAFS up to the region of 17-18 keV could be recorded in a reasonable time period using a rotating anode generator. The two generators which are most likely to be used for in-house EXAFS are the Rigaku RU-200 12 kW generator and the Elliott GX-21 15 kW generator. In Table 1, we attempted to summarize the characteristics of these two generators as they relate to EXAFS research. In addition, John Holbin and David Hempstead have contributed discussions of the GX-21 and the RU-200 generators, respectively.[z] As can be seen in Table 1 the two generators are similar in many respects. There are, however, certain differences in such areas as vacuum pumping, location of the anode assembly, low kV/high mA operation, and filament assemblies which should be evaluated for each researcher's needs prior to purchasing either unit. would like to point out that the data provided in Table 1 was obtained from the manufacturer's specifications and I urge anyone considering either of these two systems to contact recent purchasers of both generators for additional information concerning actual performance.

The second classification of x-ray sources discussed during the workshop was the focusing x-ray source such as a scanning electron microscope. Using a conventional scanning electron microscope it appears that EXAFS spectra may be obtained in the wavelength region $\sim$ 1.5-6Å on samples with $\mu_K/\mu > .1$. The use of the scanning electron microscope offers the unique advantages of focusing the spot size to less than 10 μm and scanning to obtain data from different parts of the sample. For more complete discussions of the use of the scanning electron microscopes for EXAFS please refer to the contributions to this workshop by H. W. Deckman, J. H. Dunsmuir and G. Via from Exxon and the general workshop contribution on EXAFS by Electron Energy Loss by Dale Johnson.[y]

The third type of x-ray sources discussed were the so called soft x-ray sources. Dr. J. Azoulay presented some recent work by J. Vanhatalo[3] using cold-cathode x-ray tubes to produce energies in the soft x-ray region. The use of the cold cathode tube avoids the problems of carbon and tungsten emission lines in the continuum spectrum which are encountered using conventional hot-cathode x-ray tubes. Although the work described produced

Table 1 Comparison of the RU-200 and GX-21 rotating anode
X-Ray Generators

	RU-200	GX-21
Maximum Power	12 kW	15 kW
kV ranges (Standard)[a]	20-60 kV	10-60 kV
mA ranges (Standard)	20-200 mA	10-300 mA
Be Window thickness (Standard)[b]	0.4 mm	0.25 mm
Vacuum pumping	Diffusion	turbo-molecular
Filament alignment	pre-aligned cassette	field aligned with special alignment jigs
Focal spot dimensions[c] (20 kV, 200 mA)	.5 x 10 mm	.5 x 10 mm
Column Height[d]	Fixed	Variable
Power requirements	208V, 50 Amp 3 phase	400 V, 60 Amp single phase
Vibration specifications	2μ horizontal and vertical	5μ horizontal and vertical

[a] The RU-200 is available with a factory modification to allow 5-20 kV operation.

[b] The RU-200 is available with a 0.2 mm Be Window.

[c] Focal spot sizes are variable with voltage and current settings.

[d] The GX-21 generator is available with different column height by factory order. The column height is not adjustable in the field.

count rates of only a few hundred counts per second, it does offer potential for producing photons in the soft x-ray region for in-house EXAFS. For a more detailed discussion please see the contribution to this workshop by Dr. Azoulay. The second source of soft x-rays for EXAFS discussed during the workshop was laser-produced x-rays. This was the work of Dr. R. E. Schwerzel and Dr. P. J. Mallozzi from the Battelle, Columbus labs [4,5]. They were able to record the EXAFS spectrum of aluminum foil with a single pulse of laser-produced x-rays. A 3.5 nanosecond pulse from a neodynium-doped glass laser was focused on a spot 100 to 200 μm in diameter at an intensity of about 10^{14} W/cm^2 to create a surface plasma from a metal slab. The resulting energy spectrum was then passed through the thin sample and dispersed by Bragg reflection from a flat KAP crystal and recorded on photographic film. Using this technique they can record the EXAFS spectrum for elements up to atomic No. 40, and more important, the complete EXAFS spectrum is recorded in only a few nanoseconds with a single pulse of laser-produced x-rays. This technique offers the unique possibility of measuring "flash-EXAFS" spectra of transient species having life times of only a few nanoseconds. Although commercial lasers are not available for this application, laser EXAFS does offer an exciting potential for future EXAFS experimentation. Please refer to the contribution of Drs. Schwerzel and Mallozzi for more details.

REFERENCES

[1] The advantages of synchrotron radiation for EXAFS research including the high x-ray flux and the beam polarization are well documented and were not discussed at this workshop.^{x}

[2] A comparison of sealed-tube generators and rotating anode generators for EXAFS is given by G. S. Knapp, H. Chen, and T. C. Klippert, Rev. Sci. Instrum., 49, 1658 (1978).

[3] J. Vanhatalo, L. Kaihola, and E. Suoninen, J. Phys. E9, 1156 (1976).

[4] P. J. Mallozzi, R. E. Schwerzel, H. M. Epstein, B. E. Campbell, Science, 206, 359 (1979).

[5] A. L. Robinson, Science, 205, 1239 (1979).

ROTATING ANODE X-RAY SOURCE[z]

John Holbin

Marconi-Elliott Avionics, Ltd.
Elstree Way, Borehamwood, England U.K.

A number of features of the GX-21 make it a suitable x-ray source for an in-house laboratory EXAFS facility. Chief among these are its ability to produce 300 mA beam current with voltages as low as 10 kV, experiments have even been run as low as 7 kV at lower mA settings, and the provision of a turbo-molecular vacuum pump to provide rapid evacuation of the x-ray tube.

Interchangeable anodes are available providing a large choice of target materials. Of particular interest to the EXAFS user are the high atomic number materials tungsten and gold. The careful balancing of the anodes and the rigid construction of the equipment produce broad-band vibration measurements of less than 5 micron peak-to-peak in both horizontal and vertical phases.

The entire equipment, including high voltage transformer, vacuum pumping system and recirculating water supply for anode cooling, is enclosed within a cubicle 2.316 meters long and 1.136 meters wide providing a rigid table top large enough to accomodate most experimental apparatus.

The replaceable tube filaments are easily changed and quickly aligned with the aid of jigs provided to give a range of focal spot sizes from 1 x 0.1 mm on the target. The range of adjustment provided allows for further refinement of the focal spot should the results achieved with the alignment jigs not be suitable for a particular application or operating conditions. A simple change of the tube head enables the user to select either two point sources or two vertical line sources.

An additional feature is the advanced, solid state, control system which monitors the equipment operation and protects it from damage due to variations in the power supply, coolant water flow, vacuum conditions, anode rotation speed, etc., and will restore operation to the pre-selected levels in the event of a power failure. If normal operation cannot be restored, a visual indication is given of the sub-system in need of attention.

The user has the option of selecting the x-ray port to table top height at the time of placing an order so that every configuration of experimental apparatus can be accomodated.

X-RAY SOURCE FOR EXAFS

D. G. Hempstead
Rigaku/USA, Inc., 3 Electronics Ave., Danvers, Mass. 01923

INTRODUCTION

This paper shows information presented to the sub-group on sources at the Workshop on Laboratory EXAFS Facilities April 29, 1980. Rigaku manufactures x-ray generators in a wide range of power levels. The 1.5kW, 2kW and 3kW units are used with sealed x-ray diffraction tubes. Rotating anode x-ray generators are offered in 12kW, 30kW, 60kW and 90kW capacities. This paper will present information on only the 12kW Rotating Anode X-Ray Generator, Model RU-200.

RIGAKU RU-200 (D4148 Series)

X-Ray output of this x-ray generator is 12kW maximum attained at 60kV, 200mA with a .5x10mm focal spot size. The standard unit may be operated at 20kV and 150mA with an actual focal spot size of .5x10mm on the target. With a special optional cathode, a tube current of 200mA at 20kV can be obtained with a .5x10mm actual focal spot size, a reasonable expected filament life and reduced tungsten contamination of the spectrum. Stability of kV and mA are maintained to less than $\pm 0.05\%$ for a change of $\pm 10\%$ in line voltage or $\pm 5°C$ in ambient temperature. Full output can be obtained from a 208 volt, 3 phase power line fused at 50 amps. Pre-aligned filament assemblies permit rapid change without extensive re-alignment. Anode assemblies may also be changed in less than 15 minutes with the system back at rating within an hour.

Rotating Anode Assemblies are offered with targets of Gold, Silver, Molybdenum, Copper, Nickel, Cobalt, Iron, Chromium and Aluminum to permit selection for best excitation efficiency and to avoid line interferences.

Either line or point focus can be utilized by re-positioning the Anode Assembly and changing the electron gun so the horizontal x-ray beam can be selected to match the x-ray optics. The beam height is 1200mm (4') above the floor and 300mm (1') above the table top. With the anode housing on the left, the shutter face is 50mm from the left edge of the table.

A recirculating water to water heat exchanger cools the anode, H. V. Transformer and other structures of the x-ray unit and removes the heat from the room. This is a recommended optional accessory.

An interlocked Radiation Enclosure with fail safe release, provides a barrier for scattered radiation and prevents accidental access to radiation areas. This optional accessory is recommended.

A special low kV modification can be made on an original equipment order that provides for operation at 5 to 20kV, selectable in 1kV steps. Stability is ±0.02% for kV and mA in this range.

Maximum Current at kV with Actual Focal Spot Size

200mA	20kV	0.5 x 10mm
150mA	10kV	0.5 x 10mm
100mA	5kV	0.5 x 10mm
150mA	5kV	1.5 x 10mm

The standard unit operating at rating has vibration less than 2μ in any direction measured on the anode housing or top plate.

Pre-aligned filament assemblies and cassette seal assemblies make routine maintenance easy and assure highest availability for use.

A pivot shaft located below the focal spot provides a convenient fixed pivot for EXAFS Spectrometer designs. See sketch below.

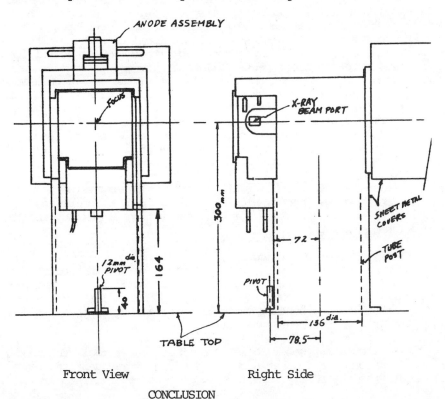

Front View   Right Side

CONCLUSION

The RU-200 X-Ray Generator is an excellent source for an EXAFS facility in a university or industrial laboratory.

USE OF A SCANNING ELECTRON MICROSCOPE AS AN
X-RAY SOURCE FOR EXAFS[W]

H. W. Deckman, J. H. Dunsmuir, and G. Via

Exxon Research and Engineering Co., Linden, NJ   07036

ABSTRACT

We have examined the possibility of performing EXAFS measurements using an electron microscope equipped with a wavelength dispersive spectrometer. Using a commercially available instrument, it appears that EXAFS spectra may be obtained in the wavelength region $\sim$ 1.5-6 Å on samples with $\dfrac{\mu_K}{\mu} > .1$, where μ_K is the K shell absorption coefficient of the element being studied and μ is the total absorption coefficient.

SUMMARY

The focused beam of a scanning electron microscope can be used to produce a point x-ray source having high emittance (> 500 w/cm^2) and low flux ($\sim$ 100 µw). Soft x-ray line emission from such a source has been used for high resolution contact radiographic analysis of laser fusion targets [1,2]. Since many scanning electron microscopes are equipped with wavelength dispersive spectrometers, we have examined the possibility of using continuum radiation from this type of source to record EXAFS spectra.

The wavelength dispersive spectrometer used in these initial experiments was manufactured by Microspec Corp. and the x-ray optics were arranged in the Johannson geometry (210 mm Rowland circle radius). Four analyzing crystals (LiF, PET, TAP and Sterate) were used to cover the spectral region from 1.14 to 92.4 Å. Resolution could be changed by crystal selection and by adjusting the size of a secondary receiving slit in front of the detector. Depending upon wavelength range a bandpass of 1-9 eV was achievable.

An AMR-1000 scanning electron microscope equipped with a LaB$_6$ gun was used to produce the focused electron beam. By selecting a 1000 µm final aperture, beam currents of 100 µA could be focused to less than a 10 µm spot size at accelerating voltages of 5-30 keV. Under these conditions, current loading on Cu, Mo, Co or W anodes (uncooled) were $\sim$100 A/cm^2. Such high current loadings were possible because of the rapid heat dissipation from the small spot. Operating a LaB$_6$ gun under conditions required to produce 50 µA beam currents, reduces filament life and also necessitates frequent cleaning of the electron optical column liner. In most cases the filament can be replaced and the column cleaned in a period of a few hours.

As expected, recorded count rates depended upon the anode material, spectrometer resolution, electron beam current and accelerating voltage. For measurements of continuum count rates in the 2.25-4 Å spectral region, all data has been scaled to reflect count rates achievable with a 50 μA electron beam accelerated to 10 kV and focused onto the surface of a W anode. For spectrometer resolutions of 7 and 15 eV, count rates of 12,000 and 35,000 cps, respectively are achievable at 3.8 Å. At 2.24 Å, a count rate of ~12,000 cps should be achievable in a 5 eV bandpass. At 1.5 Å count rates approaching 10^5 cps should be achievable in a ~ 13 eV bandpass if a 20 kV accelerating potential is used. These count rates are more than a hundred times less than those obtained using a curved crystal with a rotating anode[3,4]. At these count rates we anticipate[5] and experiment time of ~12 hrs. for samples with $\frac{\mu_K}{\mu} > .1$.

Unless the wavelength dispersive spectrometer was modified, the sample was constrained to sit between the x-ray source and analyzing crystal, rather than between the analyzing crystal and detector. This limits the type of samples which can be examined to those which do not have fluorescence lines in the region of interest, or undergo radiolysis in an intense x-ray beam. Most of the $\frac{\lambda}{2}$ component in the primary x-ray beam was removed by discriminating the output of the proportional detectors. The fluence of the source may change slightly as a function of time. To minimize this effect, the source fluence can be monitored during measurements of both I and I_0 using a separate energy dispersive spectrometer which is present in most scanning electron microscopes.

In summary we have found that EXAFS spectra could be recorded for a limited class of materials using a commercially available wavelength dispersive spectrometer and electron microscope.

REFERENCES

1. H. Deckman and J. Dunsmuir, J. Opt. Soc. Am. **69**, 1445 (1979).
2. H. Deckman and J. Dunsmuir, Proc. Topical Meeting on Inertial Confinement Fusion, Feb. 26-28, 1980 (San Diego, Ca.).
3. G. G. Cohen, D. A. Fischer, J. Colbert, and N. J. Shevchik, Rev. Sci. Instrum. **51**, 273 (1980).
4. J. A. Del Cueto and N. J. Shevchik, J. Phys. E: Sci. Instrum. **11**, 616 (1978).
5. G. S. Knapp, Haydn Chen, and T. E. Klippert, Rev. Sci. Instrum. **49**, 1658 (1978).

SOFT X-RAY SOURCES

J. Azoulay
Department of Physics, University of Washington
Seattle, Washington 98195

INTRODUCTION

Sometimes it is desired to perform EXAFS measurements utilizing soft x-ray sources. For example, then the center atom is a light element, a soft x-ray source is required to eliminate the harmonics which distort the EXAFS as follows. In the presence of harmonics the apparent absorption coefficient is given by

$$\mu_{app}(E) = -\ln \frac{I_1 \exp(-\mu_1) + I_2 \exp(-\mu_2) + \ldots}{I_1 + I_2 + \ldots} \qquad (1)$$

where $I_1, I_2, \ldots$ are the unattenuated intensities at E, 2E, ... and $\mu_1, \mu_2, \ldots$ are the absorption coefficient at E, 2E, ... A change in $\mu_{app}(E)$ as produced by the EXAFS will be given by

$$\delta\mu_{app}(E) = \frac{\delta\mu_1}{1 + \frac{I_2}{I_1} \exp(\mu_1 - \mu_2) + \ldots} . \qquad (2)$$

The measured EXAFS $\delta\mu_{app}$ is thus reduced from the correct value $\delta\mu_1$ in the presence of harmonics.

One way to eliminate this error is to eliminate the harmonics by lowering the excitation of the harmonic excitation. For low Z materials this necessitates operating the generator at low voltage. For maximizing the x-ray intensity under these conditions it is necessary to operate at as high a current as possible, not an easy accomplishment as noted in Chapter 2.

EXPERIMENTAL

X-ray tubes used today are based on Coolidge's principle. The number of electrons emitted from a heated filament is controlled independently of the potential gradient between the electrodes as described in Chapter 2. By this method stable regulation of the x-ray intensity over an extended period of time is achieved.

Because of the very low energy emission, soft x-ray generators are limited to operate without windows or with very thin windows. In both cases sophisticated and expensive vacuum systems and power supplies are required. Moreover, it is a very well known problem that carbon and tungsten contamination occurs from hot cathode which are not sealed off from the main vacuum of the instrument in x-ray spectrometers. This fact presents a serious limitation to EXAFS application.

On the other hand the cold cathode discharge tube suffers from instability in the intensity due to pressure changes in the system.

A cold cathode x-ray tube constructed by Vanhatalo et al.[2] is shown in Fig. 1. The system was evacuated by diffusion pump through a cold trap and partly open valve (throttle valve) to maintain the desired pressure level.

Using a commercial servo-operated vacuum controller to regulate the gas flow by the anode current, a stability of ± 1% over a period of several hours for CuL_α emission line intensity was obtained. This stability was comparable to that obtained with the original hot cathode x-ray tube when operated at the same power.

Aluminum L emission line spectrum obtained by the original x-ray tube (hot cathode) is shown in Fig. 2a, where considerable CK_α line shows up soon after beginning of operation. The spectrum obtained with the gas tube (Fig. 2b.) showed no measurable CK_α emission. The CK_α absorption edge seen in the spectrum is due to the 30 μ thick Formvar window.

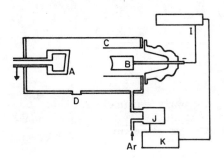

Fig. 1. A, water cooled anode. b, cylindrical aluminum cathode. c, additional electrode to reduce parasitic discharge to the walls. D, Formvar window. I, high voltage power supply. J, servo-operated valve. K, automatic pressure controller. (After Vanhatalo[2].)

These results show that gas x-ray tubes with some modifications to improve the intensity may provide a soft x-ray source for EXAFS. Also the cooling of the anode is much improved in this system as compared with the original hot cathode one.

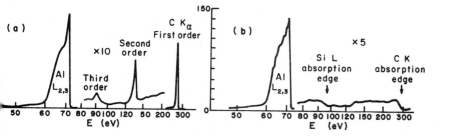

2. Al L emission spectrum taken with (a) the original vacuum tube (tungsten filament) 7 kV/50mA. (b) the gas x-ray tube 4.5mA (after Vanhatalo[2]).

REFERENCES

.S. Knapp, H. Chen, and T.C. Klippert, Rev. Sci. Instrum. 49, 658 (1978).

. Vanhatalo, L. Kaihola, and E. Suoninen, J. Phys. E 9, 1156 (1976).

LASER-EXAFS: LABORATORY EXAFS WITH A NANOSECOND PULSE OF LASER-PRODUCED X-RAYS

P. J. Mallozzi, R. E. Schwerzel, and H. M. Epstein
Battelle-Columbus Laboratories, 505 King Avenue, Columbus, OH 43201

ABSTRACT

Laser-produced x-rays are a promising alternative to synchrotron radiation for the measurement of EXAFS spectra. Experiments to date indicate that K-edge EXAFS spectra of elements with atomic numbers up to about Z=20, and L-edge spectra of elements with atomic numbers up to about Z=40, can be obtained with a single nanosecond pulse of x-rays emitted by a laser-produced plasma. The technique shows promise of providing single-shot EXAFS spectra for the remaining elements as well, with the use of advanced laser systems that are available today. The x-ray pulse can be synchronized easily with an external optical or electrical perturbation of the sample, thereby providing a unique capability for recording EXAFS spectra of highly transient species having lifetimes the order of a nanosecond.

INTRODUCTION

We have recently shown that well-resolved EXAFS spectra of light elements (e.g., aluminum and magnesium) can be obtained with a single nanosecond pulse of soft x-rays emitted by a laser-produced plasma.[1,2] In these experiments, a pulse of infrared light from a neodymium-doped glass laser is focused onto a metal target chosen on the basis of its ability to emit continuum x-rays in the vicinity of the absorption edge to be studied.

The laser pulse produces a surface plasma which serves as a point source of x-rays having a pulse width comparable to that of the laser; i.e., on the order of a nanosecond. The x-rays are passed through a light tight shield and beam-shaping slits, and are then dispersed from a flat crystal, typically composed of KAP or RAP. The basic experimental configuration is shown schematically in Figure 1. The entire sample and reference

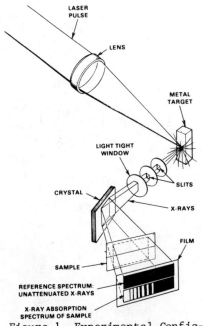

Figure 1. Experimental Configuration for Laser-EXAFS

spectra are recorded simultaneously, so that chemical structure information can be obtained from data produced with a single laser shot. We have chosen to use photographic film and computerized video densitometry for this purpose in our experiments to date, but anticipate that solid-state detector arrays will be used in the near future.

DISCUSSION

The present capabilities of the laser-EXAFS technique are illustrated by the spectrum shown below in Figure 2. This spectrum was obtained with a laser pulse of approximately 100 joules and a pulse width of approximately 3-1/2 nanoseconds focused to a 100- to 200- micrometer diameter spot on an iron slab target at an incident intensity of about 10^{14} watts/cm^2. In analyzing the data, the energy spectrum was divided into 5-electron-volt energy intervals. The number of photons that struck the film after passage through the sample was approximately 10^6 per energy interval. In principal, this allows an interval to interval contrast of $1/N^{1/2} = 10^{-3}$, or approximately 0.1 percent. This makes possible the measurement of flash-EXAFS spectra of transient species having lifetimes on the order of a nanosecond, provided the sample is concentrated. To measure flash-EXAFS spectra of the highly dilute samples that are generally of interest in biology and surface science, it will be necessary to increase the number of photons recorded per energy interval about ten to a hundredfold. It appears that this may be possible either by improving the output of continuum radiation from the laser plasma source, or by achieving better utilization of the x-rays that are already produced by employing an array of appropriately curved large crystals. Both approaches are being explored in our laboratory, and appear promising. If these experiments prove successful, flash-EXAFS studies of dilute samples will become feasible with laboratory equipment of modest cost.

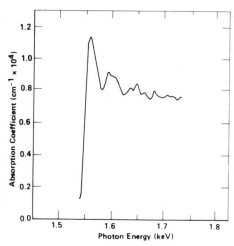

FIGURE 2. LASER-EXAFS SPECTRUM OF ALUMINUM

ACKNOWLEDGEMENT

This research was supported by the U. S. Air Force Office of Scientific Research and Battelle Memorial Institute, Corporate Technical Development.

REFERENCES

1. P. J. Mallozzi, R. E. Schwerzel, H. M. Epstein, and B. E. Campbell, Science, 206, 353 (1979).

2. P. J. Mallozzi, R. E. Schwerzel, H. M. Epstein, and B. E. Campbell, manuscript to be submitted for publication

EDITOR'S NOTES - CHAPTER 8

z. Chapter 2 by Fisher has an additional discussion of x-ray sources including some of the characteristics of the GX-21.

y. Chapter 6.

x. A comparison between laboratory and synchrotron radiation EXAFS facilities is given from various viewpoints in Chapters 7 and 12.

w. The basic manner proposed here to detect EXAFS is different than that presented in Chapter 6 even though both use the electron microscope. In Chapter 6 the energy loss spectra of the electrons record the EXAFS while here the electron beam is used to generate x-rays and the x-rays would be used to record the EXAFS as in the conventional apparatus.

WORKSHOP ON CRYSTALS AND FOCUSING

D. W. Berreman (Panel Chairman)
Bell Laboratories, Murray Hill, N. J.

INTRODUCTION

The purpose of the workshop on crystals and focusing was to exchange and pool the experience and information that the participants could contribute on the design and construction of monochromators for EXAFS with optimum speed, resolution and versatility.

The panel was chaired by D. W. Berreman of BTL and included G. G. Cohen of NBS, S. Heald of Brookhaven National Labs and Lu Kun-quan of U. of Washington. Written contributions for the workshop were made by R. J. Emrich and J. R. Katzer of U. of Delaware, R. C. Gamble of Cal. Tech., J. Crane of Cal. Tech., and T. Matsushita of S. S. R. L. G. Bunker of U. of Washington was recording secretary. G. G. Cohen and S. Heald gave invited papers on the main program that were relevant to the workshop. Other contributors were R. Hänsel of U. of Kiel, W. Germany, G. Christoff of Ohio State and D. Hempstead of Rigaku/USA Inc., J. Holben of Marconi Avionics, Hertfordshire, England and J. Hastings of Brookhaven N. L.

OBTAINING CURVED CRYSTALS

It was clear at the outset that no one design or type of crystal is optimum for all EXAFS experiments. A major problem in many cases is to procure and shape a particular kind of crystal. The chairman opened the session by encouraging EXAFS spectroscopists to participate in the shaping of their own crystal elements. Two recent books about doing optical work on crystals will be useful.[1,2] Crystalline quartz optics may be fabricated using the classical techniques for grinding, polishing and figuring glass. Care must be taken to prolong each step until the damage done by the preceding step is removed. A final light etch with HF solution will reveal any failures to remove damaged quartz, and will relieve any associated strain. Silicon and Germanium, though more brittle, may also be worked in this way.

Suppliers of curved crystal optics are few. Conferees reported great difficulty in obtaining custom-made curved crystals to test new monochromator designs. G. G. Cohen has a list of suppliers of crystal elements for X-rays in her contribution.[3] J. Holben reported sources of X-ray quality quartz crystals[4] and also of superfine polished glass for X-ray mirrors.[5] R. Hänsel reported that the Huber Co. in W. Germany will supply Ge, Si and SiO_2 perfect crystals for spectrometers of symmetric or asymmetric Johannson geometry.[6] For softer X-rays, crystals of KDP, KAP and β-alumina are useful.[7] However, they will not stand the high flux of synchrotron radiation. J. Hastings described use of tungsten-carbide layer mirrors followed by a KDP crystal to deal with synchrotron fluxes.

CRYSTAL MONOCHROMATOR GEOMETRIES[z]

Under the topic of crystal monochromator and spectrometer geometries, we had the contribution of Lu and Stern,[8] correcting a common misapprehension on the relative resolution of Johann and Johansson focusing geometries. The Johann geometry often gives adequate resolution for EXAFS and the flat crystal wafers it requires are much easier to fabricate than the Johansson cylinders. Matsushita described a spectrometer for transmission EXAFS with no moving parts that produces an entire spectrum at one time.[9] R. Hänsel spoke of the advantages of channel-cut crystals in EXAFS. They yield very high resolution and, since incident and exit beams are parallel, the detector, which may be quite cumbersome, does not need to move as far as in single-reflection devices.

It was generally agreed that double-curved crystals have not been useful for EXAFS because, up to now, nobody has made one with both high resolution and variable curvature for scanning a sufficiently broad spectrum.

BENDING CRYSTALS

A topic related to spectrometer geometry is methods of bending crystals. S. E. Crane[10] described a versatile adjustable-couple bender for elastic crystals. Crystals may also be bent against rigid backing-blocks of cylindrical or ellipsoidal or spherical form, using atmospheric pressure[11] or adhesives. With adhesives, it is particularly important that the backing block have the same coefficient of thermal expansion as the crystal. The defining surface may also be a convex frame in front of the crystal, with the crystal pressed against it.[12,13]

A simple method of fabricating surfaces for backing blocks or frames of cylindrical form invented by DuMond et al[14] is still being used with great success by G. G. Cohen and others. Cylindrical cast-iron tools for grinding glass and crystals may also be made by DuMond's method. A simple jig to prevent relative rotation of tool and crystal is also described in ref. 14.

MIRROR-CRYSTAL COMBINATIONS[z]

Curved-mirror, Flat-crystal monochromator combinations are particularly useful for work with synchrotron radiation. The high beam flux damages many crystals but the mirror eliminates much of the unneeded radiation.[15] Mirrors for X-rays have a very low acceptance angle for X-rays of short wavelength, and often also have low reflectance. Consequently, mirror-crystal combinations have not been widely used in laboratory EXAFS research. Better mirrors could possibly change this situation. Perhaps the tungsten-carbide layer mirrors described by J. Hastings will be useful there also.

Both Haensel and Katzer stressed the need for high resolution in EXAFS work. It is needed both to resolve edge structure, and to avoid washing out information from higher shells. In Katzer's opinion, 6 eV is the minimum resolution needed for useful EXAFS work at this time.[9]

Katzer[16] showed the effects that poor resolution has on fourier transforms. All EXAFS amplitudes are reduced and distorted. Amplitudes from shells at large radii are reduced more than those from shells of small radii.

G. Bunker stressed that these effects are particularly harmful when <u>absolute</u> comparisons are made in data analysis, as for example when fitting experimental data with theoretical backscattering amplitudes and phases. These effects tend to cancel if <u>relative</u> comparisons are made between unknown and model compound data measured with the same resolution, provided the distances are similar. Since the cancellation is not exact, it is clear that it is worth striving for even better resolution than 6 eV.[9]

Christoph stressed the importance of monochromator crystal surface preparation in optimizing resolution and speed. The rocking curve of the crystal should be made to match the source width (angular spread) for maximum speed. Rocking curves may be broadened by damaging the crystal surface or by bending the crystal. Matsushita also mentioned the fabrication of crystals with slight gradients in d-spacing to enhance reflection by diffusing atoms into semiconductors. This provides a method of increasing reflectivity while maintaining collimation of the beam. This method has the same effect as bending the crystal in Johann or Johansson geometries.[17]

CONCLUSIONS

There is room for improvement in speed, resolution and versatility of X-ray spectrometers for EXAFS research. No one spectrometer will cover the entire range of useful wavelengths with high speed and resolution.

Spectrometer development has been hampered by difficulty in obtaining good crystal elements of unusual materials, shapes and crystal orientations. Perhaps this is partly due to timidity on the part of experimentalists in doing their own optical work.

A reliable method for adjusting the focus of a doubly curved crystal of high resolution might be of great value in EXAFS research.

Search for crystals with large Bragg spacings that can better withstand high X-ray fluxes should continue.

Improvement in X-ray mirrors may make them more useful in EXAFS with ordinary laboratory sources.

REFERENCES

1. D. F. Horne: "Optical Production Technology" (Crane, Russak and Co., N.Y., 1972) This book contains some information on polishing crystalline quartz in addition to all sorts of information about grinding, polishing and figuring glass.
2. G. W. Fynn and W. J. A. Powell: "Cutting and Polishing of Electro-Optic Materials" (John Wiley and Sons, N.Y., 1979) This book gives information about polishing and etching a wide variety of crystalline materials including quartz, silicon, germanium, LiF and KDP.
3. G. G. Cohen: (this conference)[x]
4. Steeg and Reuter, 6 Frankfurt 56 (Nd. Eschbach), Bernstrasse 119, West Germany
5. Diano Corp., Customer Service Dept., 8 commonwealth Ave., Woburn, Mass. 01801 will supply superfine polished spectrasil glass mirrors polished to British National Physics Lab. X-ray Standards. The standard size is 60mm x 20mm x 4mm thick. Other sizes are available on special order. Gold coatings to the same standard are also available.
6. Huber Co.; American rep: Blake Industries, 660 Jerusalem Rd. Scotch Plains, N.J.
7. N. G. Alexandropolous and G. G. Cohen: Applied Spectroscopy $\underline{28}$, 155 (1974)
8. Lu Kun-quan and E. A. Stern (report follows in this chapter)
9. T. Matsushita (follows in this chapter)
10. S. E. Crane (follows in this chapter)
11. D. W. Berreman: Rev. Sci. Inst. $\underline{26}$, 1048 (1955)
12. J. W. M. DuMond, Rev. Sci. Inst. $\underline{18}$, 626 (1947)
13. Lu and Stern demonstrated a very high-resolution Johann spectrometer of germanium bent by this method at the conference
14. J. W. M. DuMond, D. A. Lind and E. R. Cohen, Rev. Sci. Inst. $\underline{18}$, 617 (1947)
15. R. C. Gamble (following in this chapter)
16. R. J. Emrich and J. R. Katzer (following in this chapter)
17. D. W. Berreman: Phys. Rev. B $\underline{19}$, 560 (1979)

JOHANN AND JOHANSSON FOCUSSING ARRANGEMENTS: ANALYTICAL ANALYSIS

Kun-quan Lu[*] and E.A. Stern, Department of Physics, University
of Washington, Seattle, Washington 98195

An analytic discussion is given of the aberrations and the relative intensities of the Johann[1] and Johansson[2] crystal arrangements. It is assumed here that the x-ray source is a point on the Rowland circle and it radiates a continuum of x-ray energies in a large enough solid angle so as to encompass the crystal. The crystal is assumed to be a perfect one with essentially 100% reflection within an angular half width $\delta \ll 1$ of the Bragg condition (half-width at half-maximum). It is found that the Johann crystal arrangement is not as inferior to the Johansson one as commonly believed.

ABERRATIONS

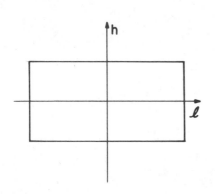

Fig. 1. The coordinate system on the crystal.

The origin of coordinates on the crystal of Fig. 1 is set at the point where the crystal planes of radius R are tangent to the Rowland circle of radius $R/2$. The length ℓ is measured along the bent crystal surface.

Considering the Johann crystal first, rays from the source to the origin make an angle θ with the crystal planes. Rays to other portions of the crystal will, in general, make different angles than θ with the crystal planes and thus will reflect other energies. The energy thus reflected is given by

$$\frac{\Delta E}{E} = \frac{1}{2R^2}\left[\frac{h^2}{\sin^2(\theta+\frac{\ell}{R})} - \frac{\ell^2 \cos^2\theta}{\sin\theta \, \sin(\theta+\frac{\ell}{R})}\right], \quad (1)$$

where ΔE is the energy difference from E which satisfies the Bragg condition at θ.

$$E_n = \frac{n\pi c\hbar}{d \sin\theta} \quad (2)$$

Here E_n is the x-ray energy of the nth harmonic where $n = 1,2,3,\ldots$, d is the crystal plane spacing, c is the velocity of light, and $\hbar$ is Planck's constant divided by 2π. We usually consider $E = E_1$, the fundamental. Note that the aberration introduced by h increases the energy while that by ℓ decreases the energy. There are thus parabolic curves on the crystal where ΔE is a constant and approximate straight lines passing through the origin where $\Delta E = 0$.

For the Johansson crystal there is no ℓ-aberration, only an aberration introduced by h.

$$\frac{\Delta E}{E} = \frac{h^2}{2R^2 \sin^2(\theta + \frac{\ell}{R})} \approx \frac{\alpha^2}{8} \quad (3)$$

where α is the angle that the ray from the source makes with the plane of the Rowland circle. It is assumed that $\alpha \ll 1$ though the first equation in (3) is true for large α.

The x-rays reflected from the crystal are focussed onto the Rowland circle at a point symmetric with the source point about the crystal origin. A slit is usually used to define the x-ray energies allowed to pass to the sample. The energies of the x-rays reflected onto the plane of the slit is next presented as a function of the coordinates shown in Figure 2. The origin of coordinates is the point of focus of the ray that reflects from the crystal origin.

For the Johann crystal

$$\frac{\Delta E}{E} \approx \frac{z^2}{8R^2 \sin^2 \theta} - \frac{1}{2\sin^{\frac{2}{3}}\theta}\left(\frac{x}{R}\operatorname{ctg}\theta\right)^{\frac{2}{3}}. \quad (4)$$

For the Johansson crystal

$$\frac{\Delta E}{E} \approx \frac{z^2}{8R^2 \sin^2 \theta}. \quad (5)$$

Note that for the Johansson crystal all rays focus along the line $x = 0$.

It should be mentioned that the ℓ-aberration presented here and the corresponding x-aberration of (4) differs from that given in Johann's original work.[1] Johann considered an extended source of a single energy. This is not the case of interest for laboratory EXAFS. Our assumption of a point source containing a continuum of energies is the more standard configuration for using the Johann focussing arrangement in EXAFS. The aberrations we find are significantly smaller on the slit plane than given by Johann.

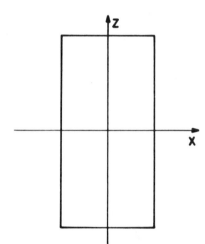

Fig. 2. The coordinate system on the slit plane perpendicular to the ray from the crystal origin.

INTENSITIES AND RESOLUTION

The relative intensities of x-rays within a given energy band reflected from the Johann, Johansson and flat crystal configuration can be estimated by calculating the area of the crystal that reflects within the rocking curve δ of the crystal about the Bragg angle θ. These results are given in the Table for a Si(400) crystal at the Cu K-edge energy assuming that $\delta = 2.5''$ of arc and h is large enough to accommodate the rocking curve. Note that the Johansson crystal is only about 1.5 - 3.4 time more intense than the Johann crystal for reasonable dimensions. Considering how more expensive and difficult it is to manufacture and align an undistorted Johansson crystal compared to a Johann crystal, it is not clear that a Johansson crystal is necessarily always the preferred one. The intensity gain in using a focussed arrangement over a plane crystal is substantial and clearly worthwhile.

TABLE

The relative x-ray intensities reflected from various configurations for a Si(400) crystal at the Cu K-edge x-ray energy. Assumed rocking curve width is 5'' and R = 20 inches.

Crystal Arrangement	Relative Intensity	
	L = H = 2 inches	L = H = 0.6 inches
Flat	3.3	1
Johann	267	180
Johansson	900	270

Next the intensity of reflection is combined with the aberrations discussed above to obtain the energy resolution. The energy resolution is defined by the usual rms formula,

$$(\Delta E)_r = \left[\overline{(\Delta E)^2} - (\overline{\Delta E})^2 \right]^{1/2}, \qquad (6)$$

where

$$\overline{(\Delta E)^2} = \frac{\int (\Delta E)^2 I(\Delta E) d(\Delta E)}{\int I(\Delta E) d(\Delta E)}, \qquad (7)$$

$$\overline{\Delta E} = \frac{\int \Delta E\, I(\Delta E) d(\Delta E)}{\int I(\Delta E) d(\Delta E)} \qquad (8)$$

Here $I(\Delta E)$ is the intensity of reflected x-rays of energy ΔE and the integration is performed over all energies admitted by the slit. The energy resolution $(\Delta E)_r$ is the rms deviation about the average value of energy $\overline{\Delta E}$. Note that $\overline{\Delta E}$ is not necessarily zero which means that the average energy is not necessarily E, the value defined by the Bragg condition. This is so because the aberrations are not in general symmetric about E.

For the Johann configuration it is found that

$$(\Delta E)_r = \frac{EL^2}{8R^2 \tan^2\theta} \left[\frac{4}{35}\left(\frac{1+K^8}{1+K^4}\right) - \frac{16}{225}\frac{(K^6-1)^2}{(K^4+1)^2} \right]^{1/2} \qquad (9)$$

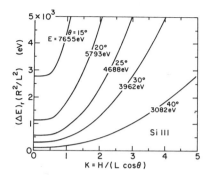

Fig. 3. A plot of the energy resolution of a Johann crystal $(\Delta E)_r R^2/L^2$ versus $K=H/(L\cos\theta)$ for $S_i(111)$ and various Bragg angles θ or x-ray energies E. $(\Delta E)_r$ is the rms resolution in eV, R is the radius of the crystal planes, L is the total length of the crystal and H is its total height.

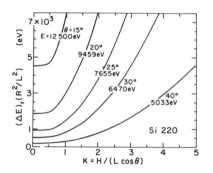

Fig. 4. The same plot as in Fig. 3 but for a Si(220) crystal.

where L is the total width of the crystal (assumed to be L/2 on either side of the origin), H is the total height (assumed to be H/2 on either side of the origin), and

$$K = \frac{H}{L \cos\theta} . \qquad (10)$$

Figures 3 - 5 show plots of $(\Delta E)_r R^2/L^2$ vs. K for Si(111), (220) and (400), respectively, as a function of θ (or E). In these plots $(\Delta E)_r$ is in eV. It is assumed that the slit collects all of the x-rays reflected from the crystal.

The resolution for the Johansson crystal is

$$(\Delta E)_r = 0.3 \frac{H^2 E}{8R^2 \sin^2\theta} . \qquad (11)$$

In both Equations (10) and (11) it is assumed that the $(\Delta E)_r \gg \Delta E_\delta$ where ΔE_δ is the energy resolution produced by the crystal rocking curve half-width δ. For Si(400) at the Cu K-edge $\Delta E_\delta \approx 0.2$ eV and for most purposes $(\Delta E)_r$ is significantly larger due to the h and ℓ aberrations as seen from Figs. 3 - 5.

CONCLUSION

There are two more contributions to the energy resolution, namely, the finite radius of the crystal planes[3] and the finite dimensions of the source.[2] This discussion focussed on the contributions to the energy resolution of the finite dimensions of the crystal because the original Johann discussion of this aberration made assumptions which are not applicable to the configuration as employed in laboratory EXAFS facilities. The corrected discussion as given here shows that the Johann configuration is not as inferior to the Johansson configuration as indicated by some investigators' interpretation of Johann's discussion.

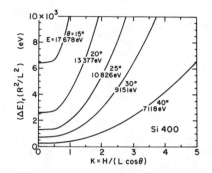

Fig. 5. The same plot as in Fig. 3 but for a Si(400) crystal.

FOOTNOTES AND REFERENCES

*Permanent address: Institute of Physics, Academia Sinica, Bejing, PRC.

1. H.H. Johann, Z. Physik, 69, 185 (1931).
2. T. Johansson, Z. Physik, 82, 507 (1933).
3. J.E. White, J. Appl. Phys. 21, 855 (1950).

X-RAY ABSORPTION SPECTROMETER WITH A DISPERSIVE AS WELL AS FOCUSING OPTICAL SYSTEM

Tadashi Matsushita*
Stanford Synchrotron Radiation Laboratory
Stanford University, Stanford, California 94305

In this note a different approach from the conventional method to the measurement of X-ray absorption spectrum is reported, where the whole spectrum of interest is measured simultaneously in a dispersive mode. In this new approach, x-ray beams converging towards a focal point are made use of, each beam of which has a continuously varying energy with the beam direction. By placing a specimen at the focal point and by measuring the spatial intensity distribution across the beam direction behind the focus with an appropriate position sensitive detector in the presence and absence of the specimen, the absorption spectrum can be measured simultaneously for a wide enough energy range.

Figure 1 shows an arrangement to realize such a geometry with a transmission type silicon crystal and a conventional laboratory X-ray source. A multiwire position sensitive proportional counter[1] with Xe gas was used. The absorption spectrum from 10μ thick Fe metal foil was measured in about 15 hours. A detailed report for this arrangement will be made elsewhere[2].

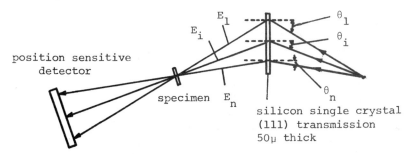

Fig.1. Dispersive method of X-ray absorption measurement with a transmission type crystal and a laboratory X-ray source.

For the case of a synchrotron radiation source, a silicon (111) curved crystal was used[3] as shown in Fig.2. The X-ray source is located far from the point on the Rowland circle. Since the glancing angles of X-rays vary along the crystal surface, X-rays of different energies are reflected at different points on the crystal surface and travel toward the focus. An X-ray film was used as the detector. Figure 3 reproduces a photodensitometer curve of a film for a Cu metal foil specimen. The exposure time was only about

---

* On leave from Department of Applied Physics, University of Tokyo.

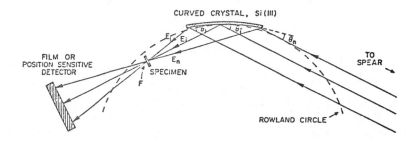

Fig.2. Dispersive method of X-ray absorption measurement with a curved crystal and a synchrotron radiation source.

0.1 sec for SPEAR operating at 3GeV and ~80mA. The energy range covered in Fig.3 is about 400eV. In a later experiment the energy range covered without rotating the crystal was extended to about 1KeV by using a larger, more uniformly bent crystal. The energy resolution was estimated at 2-3 eV experimentally. The present approach will open up the possibility of time resolved EXAFS measurement. More details will be reported elsewhere.

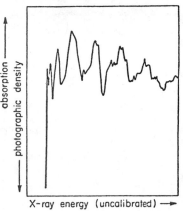

Fig.3
X-ray absorption spectrum of Cu metal foil recorded on X-ray film using a dispersive arrangement such as that shown in Figure 2.

ACKNOWLEDGEMENT

The author would like to thank Prof. A. Bienenstock for his interest and encouragement. He is also thankful to Drs. G. Brown, R. P. Phizackerley, W. Warburton and Mr. A. Filippi for their supports in various aspects. He is grateful to Dr. T. Barbee for giving him silicon single crystal wafers. Some of the materials in this work were developed at SSRL with the financial support of the National Science Foundation.

REFERENCES

1. H. Hashizume et al, Nucl. Instrum. Methods 152, 199 (1978).
2. U. Kaminaga, T. Matsushita, K. Kohra, H. Hashizume and Y. Amemiya, submitted to Jap. J. Appl. Phys.
3. T. Matsushita, SSRL Activity Report, in press.

ADJUSTABLE CRYSTAL BENDING APPARATUS

S. E. Crane
Division of Chemistry and Chemical Engineering
California Institute of Technology, Pasadena, CA 91125

The Caltech EXAFS focusing spectrometer utilizes an adjustable crystal bending mechanism similar to that described by Webb (1976)[1] and Franks (1958)[2] for focusing the x-ray beam. That is, bending moments are applied at both ends of the crystal with cylindrical couples (see Figure 1). Our system differs from those of earlier workers in that each bending element is separately adjustable, allowing translation of one or both ends of the crystal in addition to bending.

Adjustable bending systems have advantages and disadvantages compared with fixed backing plates. Among the advantages: several geometries are possible (cylindrical, logarithmic spiral) with the same bender and crystal; imperfections in the crystal grind can be corrected for after mounting; dynamic bending is possible by stepping motor control of the micrometer settings; and dust between the crystal and the backing plate is not a problem. As for disadvantages: it is not possible to achieve perfect cylindrical symmetry; it is not known how large a crystal may be bent successfully in this way; and one is never sure exactly what shape the crystal has taken.

There is a trade-off involved with the use of bending couples. The closer the bending elements of the couple are to each other, the better the approximation to a cylinder, but the greater the strain put on the crystal. Little experimental evidence exists on how severe the respective problems are. The prototype bender in our spectrometer has one-half inch of a two inch crystal between couple members (Figure 1). The portion of the crystal that extends beyond each couple is not bent, and should be minimized.

For the prototype bender, a Si (111) crystal (5.08 cm wide, 2.54 cm high, 300 microns thick parallel plate) was ground for a Johannson geometry (R=75 cm). In practice, we were able to optimize the focus of a Si (111) crystal (note that this is an elastic crystal; it is not appropriate to bend plastic crystals in this way) in a matter of a few minutes using a Cu K_α line. In one instance, however, half of the crystal was found to be reflecting the K_{α_1} line and the other half the K_{α_2} line. Obviously, some care must be taken in this regard. We were able to obtain at least a 400-fold increase in flux for a bent Johannson crystal as compared to a flat crystal of the same type.[w] For an effective source size of 50μ, the bent crystal focused 90% of the diffracted x-ray beam through a 75μ wide slit. The focus width may be further reduced by aligning the vertical tilt which has not been optimized at this time. The bandpass at 8 keV is 7 eV (FWHM).

REFERENCES

1. N. G. Webb,"Logarithmic-Spiral Focusing Monochromator,"Rev. Sci. Instr. 47, 545 (1976).
2. A. Franks, Br. J. Appl. Phys. 9, 349 (1958).

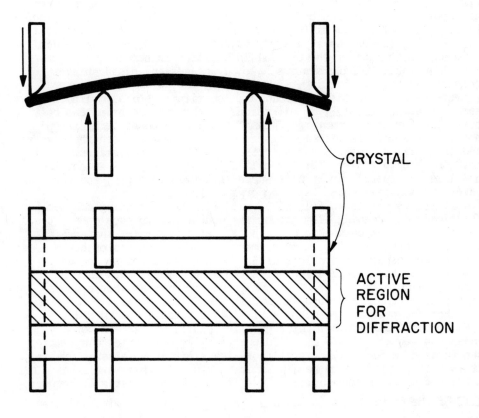

Figure 1: Crystal bending apparatus with two bending moments. The upper drawing shows the bend as viewed from the top. The arrows indicate the direction of force to induce bend. The lower drawing is a side view, showing the front surface, a portion of which must be left unobstructed for diffraction.

APPLICATION OF MIRRORS FOR FOCUSING X-RAYS [z]

R. C. Gamble
Division of Chemistry and Chemical Engineering
California Institute of Technology, Pasadena, CA 91125

Focusing x-rays from a conventional x-ray source provides one way for increasing the flux through the sample, and therefore enhancing the quality of data. Total external reflection from curved surfaces represents an alternative to bent crystals for focusing x-rays, or, more usefully, as a complement to bent crystals by focusing in a plane perpendicular to that of the crystal (Fig. 1). Such a geometry has been successfully applied for focusing synchrotron sources at various locations (for example, see Ref. 1, 2, 3, which describe a focusing system at the Stanford Synchrotron Radiation Laboratory). In addition to focusing, the angle for which reflection occurs is dependent on the photon energy[4]. This makes possible partial monochromatization of the x-ray beam. It is worthwhile, therefore, to consider the usefulness of mirrors for EXAFS spectroscopy using conventional sources.

Because the index of refraction in the x-ray region is only slightly less than unity, total external reflection only occurs at small incident angles (Fig. 2). As seen in the accompanying table, the energy dependence of θ_c makes possible the elimination of unwanted higher order reflections from the crystal, thus allowing high operating voltages of the x-ray tube. Two considerations will affect the choice of mirror material. For reasonable entrance aperatures, fairly long mirrors are required. Also, the surface roughness must be small for distances comparable to the x-ray wavelength. In the past, float or polished glass has been used. In some cases, evaporative deposition of metals has been incorporated in order to take advantage of the larger critical angles for such materials.

The focusing characteristics of the Biology beamline at the Stanford Synchrotron Radiation Laboratory provides some indication of actual performance which may be expected from mirrors. The focusing arrangement is similar to Fig. 1, where: the focusing element, F1, is a 120 cm long, ellipically curved plate of float glass; F2 is a 7 cm wide x 3 cm high flat silicon crystal, bent to a logarithmic spiral, the distance, D1 is about 1500 cm, and D2 is 150 cm. For our purposes, only the vertical focusing will be considered. The entrance aperature of the mirror is about 4 mm for 8 keV photons. The vertical focus is 250μ, resulting in geometric intensity gain of 16-fold as compared to simple aperatures defining the beam. The measured reflectivity is about 60% for a new glass plate and decreases 2- to 4-fold as debris deposits on the surface (scattering becomes significant for dirty mirrors). The overall intensity gain for 50% reflectivity is about 8-fold, an important gain for small angle scattering experiments. Also, because the synchrotron radiation diverges less than 1 mrad vertically (Gaussian distribution with about 1 cm FWHM at 1500 cm), a large fraction of the total incident radiation enters the small entrance aperature of the mirror. X-ray mirrors, therefore, are suitable for synchrotron sources.

Unfortunately, the small aperature for mirrors is not suited to the 4π divergence of conventional x-ray sources. For an EXAFS spectrometer incorporating a bent crystal, the typical Rowland circle radius, R, is 20-80 cm. The source to crystal distance may range from 10 to 30 cm, depending on the energy selected and crystal characteristics. If the source-crystal is 20 cm and a crystal of 1 cm high is used, the entrance aperature is 0.05 radians (2.9°), a value exceeding by several-fold the inclusive angle for mirrors (approximately equal to θ_c). Therefore, provided the sample may be prepared satisfactorily so as to accept the divergent beam, 1-3 cm at the sample, there is no advantage to vertical focusing. However, if the sample is small, requiring a restricted aperature, then vertical focusing may be of some benefit. The advantage of eliminating harmonic radiation, thus enabling high operating voltage on the source anode, is possible without focusing; however, the aperature angle is still restricted by the small critical angle of the reflecting surface.

It should also be noted that good focusing and reflectance requires surface smoothness from the x-ray through the optical region. Whether the mirror is ground or bent, fabricating the focusing assembly is a non trivial task. Other alternatives for increasing quality of EXAFS data should be considered carefully.

REFERENCES

1. N. G. Webb, S. Samson, R. M. Stroud, R. C. Gamble, J. D. Baldeschwieler, "A focusing monochromator for small-angle diffraction studies with synchrotron radiation", J. Appl. Cryst. 10, 104-110 (1977).
2. N. G. Webb, S. Samson, R. M. Stroud, R. C. Gamble, J. D. Baldeschwieler, "Remotely controlled mirror of variable geometry for small-angle x-ray diffraction with synchrotron radiation", Rev. Sci. Inst. 47, 836-839).
3. N. G. Webb, "Logarithmic-spiral focusing monochromator", Rev. Sci. Inst. 47, 545-547 (1976).
4. R. W. James, The Crystalline State, Vol. II: The principles of the diffraction of x-rays (Cornell, Ithaca, N.Y., 1965), 664 pp.

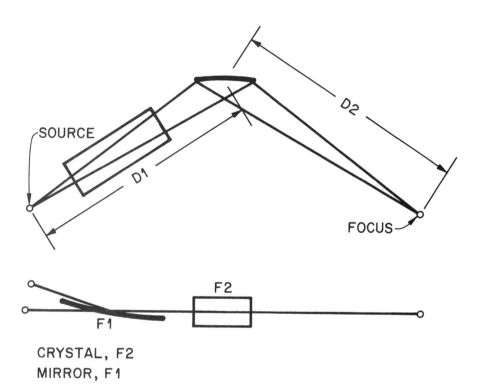

Figure 1: Ray tracing diagram showing one possible geometry for using a mirror to focus x-rays. The upper drawing is a top view showing the position of the mirror, F1, relative to the bent crystal, F2, in a Johann geometry. The side view (lower drawing) shows the position of the source relative to the in-plane position if the mirror was not present. An elliptically curved mirror provides a point focus; however, the curvature must be continually adjusted as the crystal and slits at focus are varied to maintain the Rowland circle geometry. A mirror with parabolic curvature does not require continual adjustment because the x-ray beam is, essentially, parallel in the vertical direction. In this case, the focus is a vertical line.

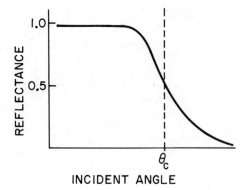

Figure 2: On the left, the reflectance as a function of incidnet angle is shown. The critical angle is θ_c. On the right the critical angle at two x-ray energies is presented for three materials which could be used as a mirror surface.

EXAFS CONFERENCE: CRYSTALS AND FOCUSING
GEOMETRY WORKSHOP

R. J. Emrich and J. R. Katzer
University of Delaware

This conference has presented several focusing geometries that increase the photon flux through the detection system and improve the energy resolution of an EXAFS spectrometer. Utilizing these ideas, one should be able to construct in the laboratory a spectrometer capable of taking EXAFS data with good statistical accuracy (< .1%) in a reasonable period of time (< 30 minutes). However, such spectrometers appear to be effective only over a limited energy range. Because of the broad energy range (~17KeV) necessary for our applications here at Delaware, it would have been very difficult to embody in single focusing geometry the simultaneous requirements of high resolution (< 1 eV), which is necessary to analyze edge structure and photon flux (~10^7-10^8/sec) necessary to record an EXAFS spectrum in a relatively short time period. Therefore, our group designed a spectrometer which uses a multicrystal monochromator system with a variable focusing arrangement. The present monochromator system uses either a double, flat crystal or a single, Johannson-cut, curved crystal to obtain monochromatic radiation.

Since it is technically easier to make high quality, flat crystals, the edge structure of our samples, most of which are catalysts, are studied using a conventional double, flat crystal monochromator (Si<311>), operating at a low ($2°$-$3°$) take-off angles and using slits when necessary to improve the resolution. This geometry allows one to obtain the resolution (< 1 eV) necessary to study EXAFS edge structure, is easily applicable to the broad range of edge energies of interest (~8 KeV-23 KeV), and can be learned about in any one of several "reference" books on x-ray spectroscopy. The main drawback of this geometry is its low intensity making it a time consuming method of acquiring data, but even this problem is not that serious since one only records a small portion of the entire EXAFS spectrum. The room temperature copper foil shown in Figure 1 is typical of the resolution possible with our instrument operating in the flat crystal mode. The calculated resolution at the copper K edge is 2.1 eV which compares very favorably with the 1 to 2 eV resolution attainable at SSRL in this energy range. The small feature on the copper K edge (arrow) is known to be about 3 eV wide. This spectrum was recorded in 12 hours but it is now possible to record a comparable spectrum in much less time.

The EXAFS portion of the spectrum is recorded using a Johannson-cut, (Si<311>) curved crystal monochromator which can be mounted interchangeably with the flat crystal monochromator. The curved crystal monochromator allows us to obtain a flux of 10^7-10^8 photons/sec which is sufficient to record an EXAFS spectrum with good statistics in about 1/2 hr. The observed energy resolution is not as good as that of the flat crystal system (6.2 eV at 8.4 eV, the Lα1, line of tungsten), but is adequate to obtain information on several coordination shells surrounding the absorbing atom. Figures 2 and 3 show transforms of a room temperature Ni foil EXAFS spectrum taken at SSRL and Delaware. The Ni spectrum taken at SSRL is particularly good having no glitches or other anomalies, and thus can be considered to be a model to compare against. The Ni spectrum taken at Delaware showed no interference from the strong emission lines of tungsten, which normally deposits on the anode during operation. These impurity lines had a peak maximum to background ratio of 10 to 1. There was, however, no evidence of these lines in the EXAFS (I_o/I) data because the response of our detectors was carefully matched. It is very difficult though, to tell how effectively these emission lines have been removed in looking at the complex fine structure. If they have not been very effectively removed, however, additional peaks or peak distortions would probably appear in the R-space transforms, i.e., the radial structure functions. Comparison in R-space should be a much more critical test of spectra quality than comparison in k-space, i.e., the EXAFS I_o/I data.

The R-space transforms of both foils show all the correct distances out to the 4th coordination shell and show no spurious, unexpected peaks. The overall magnitude of the Delaware spectrum is about 16% less than that of the SSRL spectrum. This is primarily because of the lower energy resolution of the present curved monochromator (~6 eV), as compared with that at SSRL (~1-2 eV).[V] Both spectra were collected in 15 minutes, indicating the effectiveness of the rotating anode curved crystal combination in producing quality data in a reasonably short time. Overall, the comparison between the transforms of the SSRL data and the Delaware data is very good. The decreased magnitude of the Delaware data, which affects the higher coordination shells somewhat more than the first coordination shell, is due primarily to poor energy resolution.[V] This indicates the need for higher resolution, curved crystal monochromators.

At present, we intend to obtain curved crystal monochromators with higher energy resolution, by using higher index crystal planes (Si<311>) in combination with larger Rowland radii. We are also putting in a flat crystal spectrometer on the other port of our x-ray generator and rewriting our computer software so that we may operate both a curved crystal and a flat crystal spectrometer simultaneously. This will allow us to collect 3 to 4 edge spectra a day in addition to the approximately 10 to 20 EXAFS spectra we would obtain each day if we did not have to change the monochromator system.

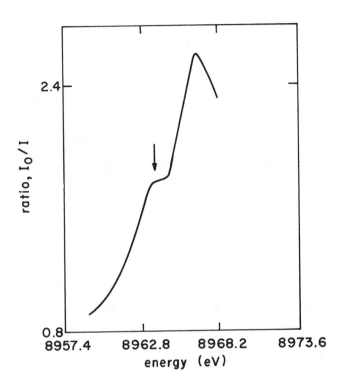

Figure 1: Copper K-edge observed at high resolution using two flat crystal system.

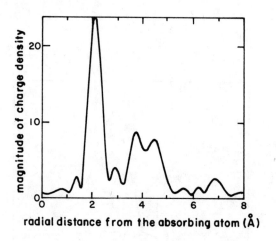

Figure 2: Radial distribution of nickel foil from Stanford Synchrotron.

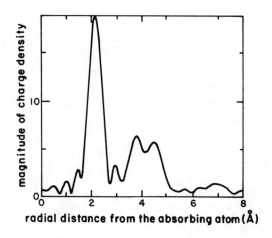

Figure 3: Radial distribution of nickel foil Delaware Line.

EDITOR'S NOTES - CHAPTER 9

z. See also Chapter 3.

y. See, however, editor's note v.

x. Not published in the Proceedings. The individual can be contacted directly for more details. Addresses are given in the Appendix.

w. This agrees well with the theoretical estimates given in Chapter 3 and in this Chapter by Lu and Stern.

v. Another possible cause of amplitude variation is the thickness effect alluded to in Chapter 6 and discussed in references 2 and 3 of that Chapter. A recent careful study of Cu metal measured with a laboratory EXAFS instrument ($\sim$5 eV resolution) and at SSRL ($\sim$2 eV resolution) showed no significant decrease in amplitude in the first shell due to different energy resolutions.* However, the thickness effect can significantly cause amplitude variations unless special care is taken. In view of this recent study it is most likely the amplitude variation found in the Emrich and Katzer contribution is due to a thickness effect. This conclusion is further supported by the fact that the higher shells of Ni are not decreased sufficiently relative to the first shell as expected from an energy resolution effect. Thus, the evidence indicates that 6 eV resolution does _not_ yet degrade significantly the EXAFS of the _first_ shell. However, further out shells are more sensitive to distortion by energy resolution effects.

*(E. A. Stern and K. Kim, submitted for publication in Phys. Rev.)

SUMMARY

D. R. Sandstrom
Washington State University, Pullman, WA 99164

Approximately a dozen conference participants attend the panel discussion on detectors, including panel members R. C. Gamble, D. R. Sandstrom (Chairman), E. A. Stern, and Y. Yacoby. Brief presentations were made by the panel members which are summarized in the accompanying reports. (E. A. Stern's contribution is included in the report of his plenary session talk on detectors.[z]) The discussion focused on the general subject areas described in the following.

A consensus was reached that harmonics and emission lines ought to be satisfactorily dealt with by appropriate design of the detector system. By achieving this goal, the constraint of operating x-ray generators below the threshold for harmonic generation would be removed, increasing available flux at the fundamental wavelength by a factor of 5-10.[y] A scenario for achieving this goal by use of a multi-channel array of detectors operated in parallel, having linearity in the neighborhood of 1% and having energy resolution sufficient to distinguish fundamental and harmonic radiation is described in more detail in the report by Y. Yacoby. An attractive attribute of such a system, which handles harmonics by subsequent correction of the data, and emission lines by high intrinsic linearity combined with generator regulation, is that it might provide a more cost effective means of upgrading a fixed anode system than the rotating anode alternative.

The option of simultaneous collection of the absorption spectrum by use of a position sensitive detector was introduced by R. C. Gamble and is elaborated upon in his report.[x] The panel considers this alternative a viable one for many applications which should be given more consideration as suitable detectors, such as self-scanning photodiode arrays, become available.

The panel discussed the sensitivity limits potentially achievable by applying the fluorescence technique on laboratory EXAFS systems. It was concluded that because of fluorescence background produced by harmonics, ultimate limitations on intensity, and differences in polarization properties of laboratory and synchrotron radiation sources (see the report by D. R. Sandstrom),[w] ultimate performance of laboratory systems is expected to fall considerably short of synchrotron source based systems.

An additional point considered was the desirability of a round-robin comparison of standard samples as additional laboratory systems become operational.

In summary, the panel members and other participants concluded that the detector system is in many respects a weak link in the present development of laboratory EXAFS systems. Optimization of detector characteristics is seen to be a relatively cost-effective and straightforward way of overcoming the limitations apparently inherent in other parts of the EXAFS system, especially x-ray generators.

EXAFS SPECTROSCOPY USING A FLAT CRYSTAL AND LINEAR DETECTOR[x]

R. C. Gamble
Division of Chemistry and Chemical Engineering
California Institute of Technology, Pasadena, CA 91125

I. The Flat Crystal Geometry

The primary emphasis of this workshop is the use of conventional x-ray sources, monochromators with focusing optics and unit-area photo detectors for obtaining EXAFS spectra. Implicit in this approach to laboratory EXAFS is that data of satisfactory quality for routine experiments, i.e., where the interesting element has sufficient concentration, can be obtained by increasing the flux of bremsstrahlung radiation through the sample. At least 500x-fold higher flux is possible with the use of a single 2.5 cm bent crystal in a Johannson geometry (R=30 cm) as compared to a flat crystal. Reasonable data may be acquired with point by point variation of energy and subsequently measuring the absorption by the sample with two detectors.

An alternative geometry is the use of a linear array of detectors or a single position-sensitive detector in which the entire EXAFS spectra is collected simultaneously from a flat crystal. The enhancement of data acquisition rate results from taking advantage of the 4π emission of x-radiation from conventional sources. For example, as shown in Fig. 1a a two-degree inclusive angle from the source onto a Si (111) crystal is sufficient to collect a 1 keV EXAFS spectrum above an 8 keV absorption edge. The linear detector must have sufficient position resolution so as to obtain 300 to 1000 data points simultaneously at a total count rate of at least 10^6/sec, and higher if a rotating anode x-ray source is used. Obviously, the detective quantum efficiency must be good; otherwise the counting statistics will suffer and consequently reduce the advantage of simultaneous data collection.

Although the overall size of the spectrometer may be adjusted to suit the detector characteristics, there are definite advantages for smaller geometry. Importantly, the loss of intensity by vertical (out of plane) divergence is reduced. In addition, smaller source size (micro-focused sources have higher brilliancy), shorter optical paths, and smaller samples all serve to increase the intensity and simplify construction.

Concerning the relative merits of the flat crystal-linear detector approach as compared to the focused geometry-unit area detector method, the primary advantages are simplicity of design and construction. Once the crystal and detector angles are fixed for a particular energy range, there are no moving parts, thus reducing the amount of precision machining and costs.

The main disadvantage is that only absorption measurements are possible. By eliminating the opportunity for fluorescence measurements, the sensitivity is limited by the detector characteristics, such as dynamic range, in addition to photon flux. Because the photons to be monochromatized pass through different parts of the sample, the homogeneity is an important consideration, and could be an important source of systematic error. Sweeping or spinning the sample will reduce this effect. Alternatively a bent crystal monochromator (not of the Rowland circle geometry) can be incorporated to reduce the effects of a divergent source. Schematically shown in Fig. 1b, the photons of different energy can be directed through only a small part of the sample, just to the rear of the slits.

In order to obtain transmission measurements normalized for the incident intensity, the linear detector must be exposed in the absence of sample. Depending on the nature of the sample and readout capability of the detector, the sample could be chopped in and out of the beam, providing a differential measurement of the transmission. Effectively, though, the data acquisition time is doubled for similar counting statistics because the incident and transmitted intensities are measured at different times. In addition, the gain and linearity of each "picture element" (pixel) of the detector must be characterized within the limits of the counting statistics.

II. The use of photodiode arrays as x-ray detectors

The flat crystal geometry places a heavy burden on the quality of the linear detector. Different kinds of position-sensitive detectors have been considered for sometime by the x-ray scattering community (c.f. reference 1 and 2 for reviews), and more recently by x-ray astronomers. However, their needs typically encompass low intensities ($<10^5$/sec) and large spatial resolution ($>100\mu$, 300-500μ typical). The flat crystal EXAFS spectrometer requires measuring higher intensities ($>10^6$/sec) at resolutions of the order 20-100μ. At these intensities, counting individual photons is unnecessary, so long as the signal representing an integrated intensity is measured in which the readout noise is small compared to the source noise.

Recent advances in photo-sensitive solid state arrays has captured our attention as possible linear x-ray detectors. In one such application[3], a silicon photodiode (SPD) array was coupled to a fluorescent film via fiber optics. Although the performance of the phototype x-ray detector was far from optimal, the test data indicated that a properly designed system based on the Reticon RL1024S could achieve a sensitivity of 10-20 x-ray photons, and a resolution of 50 microns (2 pixels) along the 2.54 cm, 1024 pixel (25μ centers) array.

Developed by Reticon (EG&G) for the mass spectrographic facility at the Jet Propulsion Laboratory, the RL1024S device has a pixel height of 2.5 mm, making it particularly suitable for spectroscopic applications. A low noise front-end amplifier and digitizer is available from Princeton Applied Research (EG&G). With suitable cooling, integration of several hours is possible.

The direct detection of x-rays with photodiode arrays has been investigated[4]; however, caution was expressed about radiation damage. Our own measurements with the RL1024S agree with the earlier calculation[4] that it is possible to fabricate a detector system for direct detection which will have a readout noise equivalent to 10 photons (8 keV) and a dynamic range of 10^4. Because the depletion zone is no greater than 10 microns, only a fraction of the incident radiation is usefully absorbed (the characteristic distance is about 70 microns for Si at 8 keV), and therefore the detective quantum efficiency exhibits a strong energy dependence within this region.

The difficulty of a shallow depletion region has been circumvented by fabricating charge-coupled devices on high resistivity silicon substrates by researchers from The Advanced Technology Laboratory, Westinghouse Corp., Baltimore[5,6]. Depletion layers of 150-250 microns make possible large active sensing volumes and therefore high detective quantum efficiency for x-rays (>80% from 0.5 to 15 keV). Although still in its infancy, this technology will most likely provide the linear detector characteristics which are required for the flat crystal geometry.

A commercial spectrometer based on the flat crystal geometry would be useful for research and development, where compounds can be analyzed which contain a large proportion of the interesting element, and for industrial control where EXAFS spectroscopy may conveniently monitor the quality of manufactured materials. Although interpretation of EXAFS spectra is still a difficult science, the data may be used as a "fingerprint" of a material which has certain desired properties that is dependent on local structure about the element of interest.

REFERENCES

1. H. Winick and G. Brown, "Workshop on x-ray instrumentation for synchrotron research", SSRL Rept. #78/04, May 1978, Stanford Synchrotron Radiation Laboratory, Stanford University, Stanford, CA.
2. J. Scheten and R. W. Hendricks, "Recent developments in x-ray and neutron small-angle scattering instrumentation and data analysis", J. Appl. Cryst. 11, 297-324 (1978).
3. R. C. Gamble, J. D. Baldeschwieler and C. E. Giffin, "Linear photo-sensitive x-ray detector incorporating a self-scanning photodiode array", Rev. Sci. Instrum. 50, 1416-1420 (1979).
4. "Application of Reticon photodiode arrays as Electron and x-ray detectors", Applications note #101, Reticon Corp., Sunnyvale, CA 94086 (1975).
5. M. C. Peckerar, D. McCann, F. Blaha, W. Mend and R. Fulton, "Deep depletion charge-coupled devices for x-ray and IR sensing applications", International Electron Devices Meeting, by IEEE, Washington, D. C., Dec. 1979.
6. D. M. McCann, M. E. Peckerar, W. Mend, D. A. Schwartz, R. E. Griffiths, G. Polucci and M. V. Zombeck, "Area array x-ray sensors", Proc of SPIE, vol. 217-15, in press (1980).

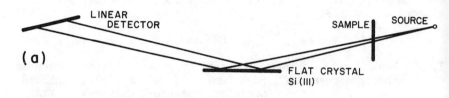

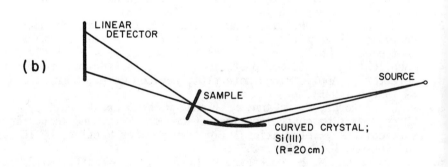

Figure 1: Ray tracing diagram for two possible geometries using a linear detector and a flat (1a) or curved crystal (1b). Depending on the spacial resolution of the detector, other crystals which allow larger Bragg angles may be more suitable.

POLARIZATION CONSIDERATIONS IN THE FLUORESCENCE DETECTION OF EXAFS

D. R. Sandstrom and J. M. Fine
Washington State University, Pullman, WA 99164

Scattered radiation produces the background signal in the fluorescence detection of EXAFS. The degree of polarization of the incident x-rays and the placement of the detector relative to the plane of polarization significantly affect the magnitude of this signal. We have calculated this effect for a measurement configuration approximating a practical EXAFS experiment for the cases of 100% plane polarized and unpolarized incident radiation. The results are significant for laboratory EXAFS measurements, in which the polarization properties may differ significantly from those of synchrotron radiation.

Because the energy spectrum of the scattered radiation lies higher than the fluorescence line, detection schemes typically discriminate in energy or wavelength, or partially suppress the scattered radiation using filters.[1] Even if these techniques are exploited, a reduction in the intensity of the scattering into the detector will ultimately lower the sensitivity limit of any experiment.

Calculations of the scattered intensity have been carried out for a configuration consisting of a dilute solution of absorbing atoms in water. The sample volume is of infinite thickness with a planar surface set at 45° to the incident beam direction. The detector center is set at 90° to the incident beam and 45° to the sample surface. The detector area is a rectangular patch on a spherical surface centered on the sample. The patch is defined by parallels (constant scattering angle relative to the beam) and meridians (constant azmuthal angle about the beam) on the spherical surface, drawn symmetrically about the detector center, which is the polarization direction in the case of polarized radiation. In the present approximation, electron binding effects have been neglected for the low Z scatterers and the Klein-Nishina formulas[2] were used to calculate the differential scattering cross sections per electron. An integration was carried over the detector surface which took into account the variation of scattered x-ray energy with scattering angle, variation of sample absorption coefficient with energy, and differing escape distances to different portions of the detector surface.

Figure 1 shows the ratio of the scattered intensity for unpolarized versus polarized incident radiation. The large ratio for small detector size reflects the absence of scattering along the polarization direction. This phenomenon provides a significant advantage for practical detectors (area ~ 1/10 of 4π) for which polarized radiation will produce 3-4 times lower scattering background than unpolarized radiation. Also, when the incident radiation is partially or fully polarized, differences of at least this magnitude can be expected to depend on detector placement.

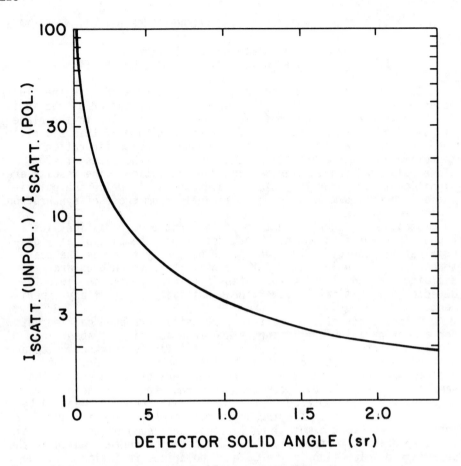

Fig. 1. Ratio of scattered radiation intensity for polarized vs unpolarized incident radiation for water scatterer and incident beam energy just above the Ni-K edge (8333 eV). See text for sample and detector configuration; sample is centered on polarization direction for the case of polarized incident beam.

REFERENCES

1. E. A. Stern and S. H. Heald, Rev. Sci. Instrum. 50, 1579 (1979).

2. R. D. Evans, "Compton Effect," in Corpuscles and Radiation in Matter II," S. Fluegge, Ed., Encyclopedia of Physics (Springer, Berlin, 1958), pp. 218-97. Differential scattering cross sections for polarized and unpolarized radiation are given by Equations 13.5 and 15.2 respectively.

OPTIMIZATION OF THE X RAY DETECTION SYSTEM FOR EXAFS

Y. Yacoby
Racah Institute of Physics, Hebrew University, Jerusalem, Israel.

We propose that an inexpensive way to upgrade an EXAFS system is to optimise the detection system. The useful X-ray intensity can, in this way, be increased by a factor of up to 10.

Harmonics in the X-ray output of the monochromator produce distortion and noise.[v] In present Laboratory systems these are avoided by operating the X-ray generator at a voltage too low to produce the harmonics. This procedure has three adverse effects. First, the X-ray tube cannot be operated at full current leading to a loss in beam intensity. Second the cathode's life time shortens. Third the intensity of the fundamental X-ray beam is lower than what can be obtained if the generator is operated at a higher voltage by a factor [1]:

$$v(1-v_\lambda/v)$$

where v is the voltage of the X-ray generator, v_λ is the corresponding threshold voltage for wavelength λ.

Harmonics present less of a problem in luminescence experiments because their only adverse effect is to raise the background and produce a higher noise level. Calculations show that in this case working with harmonics at a higher X-ray generator voltage improves signal to noise ratio.

In the absorption case we propose to use the detection system to discriminate against the harmonics. This detection scheme should meet two requirements:
a. It should resolve the energy of the X-ray photons and discriminate electronically against the harmonics.
b. It should resolve and count at a rate of up to 2×10^6 pps with deviations from linearity that do not exceed 10%. This count rate incorporates the increase due to the higher generator voltage, the contribution of the harmonics and the absorption of the sample.

This required count rate cannot be achieved by a single channel of amplifier and pulse height discriminator. However the gain involved in the system's performance warrants the use of about 10 parallel electronic channels.

Both multi wire proportional counters[2] and scintillators[3] have sufficient resolution to distinguish between the fundamental and the harmonic X-ray photons.

We may consider two detector schemes[u]

a. Multiwire proportional counter
   The detector should have a sufficient density of positively biased wires to allow short enough photo electron flight times. The detector should be placed where the X ray radiation density is sufficiently low so that the probability that two photons will produce overlaping pulses in the same wire will be very small. Groups of wires should be connected to the same amplifier such that the count rate in the group will not exceed 10^5 sec^{-1}. Each such group of wires should be electrically separated from the others to avoid cross talk among the channels.
   The proportional counter detection scheme has the advantage that it can be used to measure both the reference I_o and the signal I, in the EXAFS experiment. However it has the disadvantage that it requires different gas mixtures for different X ray photon energy ranges in order to optimise quantum efficiency.

b. Scintillator
   Scintillators have the advantage that they have 100% quantum efficiency over a wide range of energies. Their resolution on the other hand is about 35% which in this case would be marginal. The scintillator crystal should be placed at a relatively large angle with respect to the X ray beam so that the radiation density be small and the crystal used may be thin. The crystal should be cut into 10 parts each separately aluminized so that the optical photons produced by each X ray photon are confined to a well defined area. The optical photon should then be amplified by an image intensifier plate followed by an array of optical diode detectors.
   The main drawback of this scheme is that it cannot be used to measure I_o, thus making it necessary to operate in the "sample in sample out" mode.

REFERENCES

1. Eugene P. Bertine, principles and practice of X-ray spectrometric analysis (Plenum Press New York), p. 11
2. Ibid, p. 162, p. 173
3. Ibid, p. 178

EXAFS CONFERENCE: DETECTION SYSTEMS

R. J. Emrich and J. R. Katzer
University of Delaware, Newark, DE 19711

A number of different detection systems for EXAFS experiments have been discussed at this conference and in this workshop. It has generally been agreed upon that improving the detector system is the most cost effective way of improving the EXAFS experiments. To be effective in a laboratory transmission EXAFS experiment, the detector system must have sufficient resolution to distinguish between the x-ray radiation of interest and the first harmonic radiation which is also produced by the monochromator system. It is also important that the detector system be highly linear over a dynamic range of 100 to 1 and have a high efficiency with respect to the energy of interest. The system that has been devised at Delaware meets these requirements, is fairly simple, and yet very effective. Also, this system requires no modification of the x-ray source to reduce photon flux when one scans through any emission lines which may occur in the energy range of interest.

The detection system at Delaware uses gas flow detectors to measure the incident (I_0) and transmitted (I) intensities. The detectors are right circular cylinders with 0.001" thick windows of berylium foil situated symmetrically opposite along the length. The I_0 detector is the smaller of the two (2.5 cm O.D. 7.2 cm length). This permits a larger fraction of the incident photons to pass through it; it was designed to absorb 10 to 15% of the photons at 8.9 keV, the copper K-shell absorption edge. The I detector is designed to collect almost all the photons that have passed through the sample, which ideally is between 10 and 30% of the incident radiation, therefore, it is somewhat larger (5.0 cm OD. 14.7 cm length). P10 (90% Ar, 10% CH_4) gas flows through both detectors at approximately 40 cc/min.

P10 gas was chosen for the detection system to reduce the sensitivity of the detection system to the third harmonic ($\lambda/3$) of the Si <311> monochromator crystal. In diffraction, the structure factor is zero for $\lambda/2n$ for all odd number planes of Si. Whereas $\lambda/2n+1$ radiation has non-zero structure factors. Thus there is no second harmonic only a third harmonic. P10 gas has a very high quantum efficiency in the neighborhood of the Cu K-edge, but its quantum efficiency drops markedly for the third harmonic. This judicious choice of monochromator crystals thus allows us to run our source at a higher voltage than if the second harmonic were present which in turn allows us to operate the source at a higher current and achieve higher spectral flux. We therefore have high flux and no interference from higher harmonics as evidenced by the nickel foil transforms shown at the crystals and focusing workshop.[t] Other fill gases are used when the spectrometer is operated in a significantly higher or lower energy range.

These detectors may also be operated either as proportional counters or as ionization chambers. As proportional counters, the signals may be put through pulse height analysis to remove harmonic radiation if necessary. Operation in the proportional counter mode is usually done when the double-flat crystal monochromator is used. Edge structure is studied in this mode of operation: the photon flux is relatively low; collection times are long, on the order of hours; but the resolution is good, approximately 2eV. However, the linearity of the proportional counter signal degrades significantly at high fluxes (>10^6 c.p.s.).[u] This precludes using the detection system in this mode when the Johannson-cut, curved crystal monochromator is installed in the spectrometer as it is when EXAFS spectra are being collected. Under such circumstances, the detectors are operated as ionization chambers. Data glitches which are due to non-linear behavior of the detection system when the input flux goes through large increases (i.e., characteristic emission lines) or decreases (i.e., added monochromator reflections) has been eliminated by careful matching of the detectors and thus there is no need to modify the x-ray source.[u]

EDITOR'S NOTES - CHAPTER 10

z. Chapter 4.

y. See Chapter 2 and the contribution of Y. Yacoby in this Chapter.

x. See also contribution of T. Matsushita, Chapter 9.

w. Contribution by D. R. Sandstrom and J. M. Fine in this Chapter.

v. See Chapter 1.

u. See also Chapter 4.

t. Emrich and Katzer contribution in Chapter 9.

WORKSHOP ON AUTOMATING AN EXAFS FACILITY: HARDWARE AND SOFTWARE CONSIDERATIONS[z]

P. Georgopoulos*
Materials Science Division, Argonne National Laboratory
Argonne, IL 60439

D. E. Sayers
Dept. of Physics, North Carolina State University at Raleigh
Raleigh, NC 27650

B. Bunker and T. Elam
Dept. of Physics, University of Washington
Seattle, WA 98195

and

W. A. Grote
Physical Sciences Center, Monsanto Corporation
St. Louis, MO 63166

ABSTRACT

The basic design considerations for computer hardware and software are reviewed, applicable not only to laboratory EXAFS facilities, but also to synchrotron installations. Uniformity and standardization of both hardware configurations, as well as program packages for data collection and analysis, are heavily emphasized. Specific recommendations are made with respect to choice of computers, peripherals and interfaces, and also guidelines for the development of software packages. A description of two working computer-interfaced EXAFS facilities is presented which can serve as prototypes for future developments.

INTRODUCTION

The need for some kind of automation of EXAFS experiments, as well as computerized data reduction and analysis seems hardly worth emphasizing. Each measurement produces large amounts of data which, in the interests of large throughput and efficiency, must be rapidly retrieved, stored and analyzed. It became, however, apparent quite early in our discussion that a capability for on-line data analysis was also highly desirable (some of us felt it was absolutely necessary). Immediately after a measurement, one should be able to carry out at least a preliminary analysis far enough to assess the quality of the data and detect the presence of unwanted effects inherent in an EXAFS

---

*Work supported by the U. S. Department of Energy.

experiment (harmonic contamination, pinholes in the sample, sample deterioration, glitches, etc.).

Having agreed upon this requirement, the hardware configurations that we considered obviously had to include either a mini-computer or a microprocessor. As we shall see below, such an arrangement makes possible certain other features that were deemed convenient or necessary (such as flexible graphics capabilities, easily transportable storage media of general acceptance etc.).

Within this framework, the Workshop then sought to accomplish the following goals:

1. To identify specific features that the hardware configuration must incorporate.

2. To formulate recommendations about specific hardware components, based on their cost, capabilities and general acceptance by the scientific community.

3. To define the functions that the software package must be capable of for efficient and rapid data collection, evaluation and analysis.

4. To outline a general philosophy to be followed in the development of software that is powerful, flexible and easy to use and transport from one EXAFS facility to another.

The underlying trend throughout the discussion was toward uniformity and standardization. It was felt that with the expected large increase in the number of EXAFS facilities, both in various laboratories and at synchrotron installations, it would be extremely important that a general software package exist directly compatible (or at most, after minor and easy to implement modifications) with all hardware configurations, well documented and easy to use by the average experimenter with some familiarity with small computers. This seems to place quite severe restrictions on the choice of hardware, but with careful design of the software and given the significant computing power of today's mini- and micro-computers, considerable latitude in the hardware configuration is possible. This will be made quite clear in the following two sections.

The conclusions and recommendations of this Workshop are not specific only to laboratory EXAFS facilities, but apply equally well to synchrotron installations also. In fact, there was a conscious effort to take into account in our discussions the particular requirements of the latter (higher emphasis on large throughput, users with somewhat more restricted experience with EXAFS instrumentation and analytical methods, etc.). This picture is expected to change with the proliferation of laboratory facilities and the availability of many more dedicated beam lines at the new National Synchrotron Light Source at Brookhaven National Laboratory. However, the need for standardization will become even more pronounced. The mode of coexistence of laboratory and synchrotron facilities, as perceived by this Workshop, will be such that simple experiments, not requiring extreme photon fluxes, as well as exploratory experiments, will be performed at laboratory facilities in the interests of convenience and flexibility, whereas the synchrotron facilities will be reserved for sophisticated, well planned experiments.[y] It will be extremely important under these circumstances that the synchrotron user find essentially the same instrumentation as in his own laboratory and the same software that he is familiar with. In addition, he should be able to carry his data back to his laboratory on a storage medium and under a format entirely compatible with his computer.

In the following sections we will examine in detail the various aspects of hardware and software configurations that the Workshop found appropriate. In addition, we will describe two already working computer interfaced EXAFS facilities, at Argonne National Laboratory and Monsanto Corporation, which serve to illustrate the application of some of the ideas expressed at this Workshop.[x]

COMPUTER HARDWARE

The computer hardware discussed by the Workshop included the central computer, which is generally a mini-computer or microprocessor based system, the peripherals needed for data storage, graphics display of data and communication with other computers, and the interface to the electronics which take data, control the spectrometer and monitor or control other related instruments and parameters (such as sample temperature, generator current, etc.).

The general discussion of hardware attempted to define the central or critical features to be included, describe possible hardware configurations and the range of costs involved and examine what standardization has occurred or is possible. To begin a more detailed description of the hardware, the typical configuration was broken into components including:

1. The control computer and terminal.
2. Program and data storage capability.
3. (Optional) The interface with another computer for analysis or data storage.
4. (Optional) A graphics terminal and,
5. The interface to the electronics.

1. Central computer

In almost all cases this will be a mini-computer or a microprocessor based system. With the possible exception of operating in a multi-user environment, the requirements of an EXAFS experiment are easily met by many computer systems now on the market. The only specific points that were raised by the Workshop were that at least 64 kbytes of memory were needed and that smaller amateur or home-based systems should be avoided, if possible, because of difficulty with interfacing and lack of software support. A survey of existing and planned systems demonstrated that a de facto standardization has already taken place, in that all but two systems were based on DEC 11 series mini-computers or LSI-11 microprocessors, running under the RT-11 operating system with foreground/background capability or the RSX-11M time-sharing operating system. Since this is also compatible with the existing and planned systems at SSRL and CHESS and with most, if not all, systems proposed for the National Synchrotron Light Source, and since, as mentioned earlier, many users will be using both a laboratory EXAFS facility, as well as the synchrotron radiation facilities, this seems to strongly favor standardization to these computers.

2. Data and program storage

The minimum storage which was recommended was a dual floppy disk drive, in which programs were stored on one diskette and data on the other. An additional advantage of the floppy disks is that they are relatively inexpensive and a convenient way to transfer data and programs from one system to another. The need for additional storage capacity will be dictated by the particular computer configuration available to each user. If no other computer is available or the user chooses to perform most of the analysis on the same computer which controls the experiment, then a substantial amount of mass storage (e.g., a dual hard disk drive) will be needed. It is also likely that many groups will operate in an intermediate mode in which some data analysis is done on the local computer and then transferred to a larger computer for the final analysis and long term storage. In this case, the storage capacity of dual floppy disks will probably be sufficient (a single hard disk or a Winchester disk drive may be more appropriate in certain cases).

3. The interface with another computer

If a larger computer facility is available and easily accessible, then a high speed link for data transfer, while optional, is highly desirable, since it allows the user the option of data analysis and storage either on the local computer or at the larger facility. If, however, a large computer is not available or if only a slow link (e.g., 110 baud) is possible, then the local computer must have increased capacity, both in terms of memory and mass storage, as discussed above.

4. Graphics terminal

The Workshop also felt that this was an optional but highly desirable capability to have. The ability to do enough data analysis in an interactive mode to assess the general quality of the data in terms of signal-to-noise ratio, interference by characteristic lines or other systematic errors is extremely important, if the full capability of the EXAFS technique is to be utilized. Whether the analysis is done on the local computer or the larger installation was again dependent on the particular facilities available to the user, provided the preliminary analysis could be performed shortly after the experiment. The Workshop did seem to prefer the Tektronix 4000 series terminals and recommended that, if possible, the graphics terminal be separate from the control terminal of the computer. A hard copy facility with the graphics was also desirable.

5. The interface to the electronics

The survey of existing and planned EXAFS installations showed that there was a uniform mix of homemade interfaces, CAMAC and NIM, with the particular choice determined by a variety of factors, including cost, familiarity and available manpower. It was generally felt that CAMAC should be the best system for standardization since it is relatively easy to use and alter, and it is also being used extensively at the national synchrotron radiation facilities. Among its other advantages is that, by its concept, it allows easy interfacing to a variety of instruments, such as stepping motors, scalers, digital voltmeters, etc. On the minus side is the fact that its cost is considerably higher than any alternative solution.

In summary, it can be stated that the general hardware needs for an EXAFS facility seem to be well within the current technology. There is a wide variety of possible hardware configurations, with the particular choice being determined by a combination of cost, background and preference of the user and the expertise of the available manpower to develop the facility. The costs for the hardware can range from $5,000 to $50,000 with a typical cost between $20,000 and $30,000. Standardization of both the hardware and the software, not only among the laboratory EXAFS community, but also with the national synchrotron radiation centers is highly desirable. The hardware components which the Workshop recommended to be used for standardization included DEC 11 series mini-computers or LSI-11 microprocessors, running under RT-11 or RSX-11M operating systems, and CAMAC electronics. Some form of graphics capability was found to be highly

desirable. Data and program transfers between facilities should
be most effectively carried out through floppy disks.

SOFTWARE DESIGN

There was general agreement throughout our discussion that
proper design of the software package can allow considerable
latitude in the choice of hardware components. This can be
accomplished by adopting and adhering to a certain programming
philosophy to be expanded below, with regards to program
architecture and general layout. Here are certain design goals,
which programmers must always keep in mind:

1. "Top-down" design, emphasizing convenience to the user. This
can be best accomplished with an interactive, conversational
program, which prompts for user input, anticipates certain user
responses and is fool-proof with respect to the safety of the
spectrometer and integrity of the data files in storage.

2. Relative machine independence. This can be best achieved by
isolating all machine-dependent codes in low-level subroutines to
perform primitive functions. An example of this would be moving
the monochromator a prescribed number of steps. Different systems
will obviously require quite different code to implement this, but
the operation should be totally transparent to the user, i.e., the
same command should lead to the same end result, irrespective of
the hardware configuration. With this type of linkage uniformity,
the great bulk of the program will be essentially unchanged when
transporting from one facility to another, plus the user will feel
at home in any facility. This is especially important to synchrotron users who also have an EXAFS system in their laboratory.

3. Modularization. This is related to paragraph 2 above, but is
broader in scope. The basic thrust here is to follow structured
programming practices, making programs more readable and program
maintenance and modification easier. Ideally, the main program
should be a simple command cracker, branching to subroutines for
the actual execution of commands.

4. Avoidance of any programming "tricks" to get around standard
FORTRAN IV, such as overindexing arrays, renaming common
variables, using machine-dependent subroutine calling schemes,
etc. Many of us have used these tricks in the past, thinking that
the gained efficiency would be worth the confusion. In practice,
the code becomes almost unreadable and unverifiable, even by the
programmer himself after a few months. If it becomes very
desirable to resort to such schemes, they should at least be
isolated in well-annotated subroutines which can be treated as
"black boxes".

5. The program should be "soft", making all hardware addresses, crystal parameters, etc. easily changeable by the user. Additionally, the user should be able to customize his own default parameters in an initialization routine. There may or may not be master defaults in this scheme. The idea is to make the code flexible enough to be convenient to all users and to discourage tinkering with the source code, unless absolutely necessary.

6. The format of all input and output files must be identical to ensure portability of data from one facility to another. This again, is very important for synchrotron users, who also have EXAFS apparatus in their lab.

Certain means and ways to achieve these goals are sketched below. This is not a complete list of programming practices to be followed and implemented, but is does illustrate the general programming philosophy.

Notes on human engineering:

The concensus of this Workshop was that the program to be developed should be conversational, prompting the user for inputs. This is certainly the best solution for a beginner user, but it may be too slow for someone with experience. The optimum solution (implemented in certain Operating Systems) is to prompt for input only if the necessary parameters are not given on the command line. An experienced user can then type ahead without waiting for all the prompts. This would be especially convenient on a hard-copy terminal.

--- Reasonable defaults should be implemented so that even a half-alert user is safe from grave consequences. For instance, when unreasonable parameters are entered, the program should re-prompt with a warning message. Input parameters for scans should be checked for consistency, disk directories should be searched to ensure that an earlier set of data will not be overwritten, etc. The basic goal is to make the system as foolproof as possible, without having the user fight it all the time.

--- If possible, scan parameters should be storable for easy access from within the program. Many scan setups can be quite complicated, with more than four or five different step sizes and integration times over various ranges of the scan. It takes long enough to input these numbers that they should be easy to store and recall.

--- An on-line HELP facility should be considered. If it is properly designed to give the user the right amount of information, it can be very helpful, even with a user's guide nearby (obviously, one should be able to use HELP without consulting the user's guide!).

--- All MOVE commands, scan parameter input, etc., should be expressible in either monochromator steps, angle or energy units. To toggle the mode in, a MODE command should be implemented with local override capability; for instance, if the mode is currently degrees, an input specification "MOVE 9300 EV" should move the spectrometer to a setting appropriate to an energy of 9300 eV and not to an angle of 9300 degrees.

In summary, the human engineering facet of programming is to second-guess the user's needs and ways of getting in trouble. If most of his problems can be anticipated (for instance, by making sure that all commands are clearly distinct and do not look similar), the likelihood of error should be greatly reduced. Most of these ideas seem pretty much common sense, but we decided to state them explicitly, as they should be always kept in mind by the programmer.

Notes on program architecture:

The program should be modular, with the main segment being a simple command cracker. Control should pass to the appropriate subroutine for the actual command execution. A number of low-level utility subroutines should handle all apparatus control and all machine dependence should be well isolated. With this type of modularization, the resulting program will be much more readable and easy to update. Overlaying the program will likewise be quite straightforward. The isolation of all machine dependence in well defined segments will ensure portability among EXAFS facilities.

The modularization of code into a command cracker and called subroutines has one additional and very important advantage, in light of the fact that the software package will probably be developed as a collective effort by many scientists: In principle, all that will be necessary to add a new function amounts to the addition of a few statements to the main program and the linkage of the new subroutines, with no (or, at worst, minor) modification of the existing code. In addition, any changes and upgrading can be documented quite conveniently.

THE ARGONNE EXAFS FACILITY

The Argonne National Laboratory EXAFS facility, in its current state, is the culmination of four years of intensive development. During this period the facility has undergone many changes and upgrades and much has been learned about its operation and ways of improving its performance. Here, we will only concentrate on the automation and data collection and analysis software developed since the facility became operational.

The initial spectrometer was interfaced to an INTEL IMSAI 8080 microprocessor equipped with a single floppy disk drive and a

Tektronix 2648A graphics terminal with a 4632 hard copy unit. The
interface included provisions to operate a stepping motor
controller, service a scaler and actuate a pneumatic cylinder to
move the sample in and out of the x-ray beam.

The major drawback of this system was the almost complete
lack of software support. The Disk Operating System was very
primitive and the only utility programs available were a
Peripheral Interchange Program (PIP), an Image Transfer Program
between floppy disk and memory, an Assembler and a stripped-down
version of a FORTRAN IV compiler without any library functions.
Therefore, other than straight data acquisition and minor
corrections, no further data analysis could be carried out on-
line. The data file had to be transferred to the central IBM
370/175 computer via a slow (110 baud) link for processing.

When a second EXAFS spectrometer was installed on a newly
aquired rotating anode x-ray generator, a second automation system
was put together, much more primitive than the first. This one
employed a multichannel scaler and a number of pulse generators
properly daisy-chained to provide pulses for the stepping motor
and MCS advance. At the end of each run, the MCS memory was
dumped on cards and subsequently read into the IBM 370/175 for
processing. The shortcomings of this mode of data acquisition
became apparent very soon and it was decided that a major
improvement was necessary in the interests of adequate throughput,
convenience and reliability.

The current EXAFS facility employs a DEC LSI-11
microprocessor with 64 kbytes of memory, a dual 5 Mbyte hard disk
drive, paper tape reader punch, line printer, X-Y scope and
plotter. A general purpose interface was built, featuring a 16
channel, 12 bit multiplexed A/D converter, an 8 channel, 12 bit
D/A converter, 16 TTL input and output ports with interrupt
capability and two scaler service ports (the scalers can be reset,
started, stopped, read, etc.), also with interrupt capability.
This interface is currently connected to drive the stepping motor
controller, service the two scalers, operate the X-Y scope and
plotter and adjust the current of the x-ray generator and the set
point of the sample cryostat. Many additional functions can be
accomplished through this interface, as they become necessary,
such as added stepping motor controls, safety interlocks, etc.

Perhaps the most important advantage of this new system is
the powerful RT-11 foreground/background Operating System, with a
comprehensive set of sophisticated utility programs. Particularly
valuable is the FORTRAN package available, with an extensive
arithmetic and utility library, overlay and interrupt capability
and the possibility of intermingling FORTRAN and assembly language
subroutines in the same program. FORTRAN, rather than BASIC, was
chosen, being a much more sophisticated computational language and

one which everybody is familiar with to a lesser or greater
degree.

A fairly comprehensive EXAFS software package has been
developed for data acquisition and analysis. Its architecture
complies quite closely with the programming guidelines set forth
in the preceding sections of this report, hence it can serve as a
starting point for future development of a universal EXAFS
software package.

The program consists of a root segment and called
subroutines which can overlay each other, since the vast majority
of them do not have to be coresident in core. This scheme affords
unlimited expansion capabilities. The root segment (main program)
is a simple command cracker and skip chain that passes control to
the appropriate subroutine. Figure 1 shows a schematic of the
operation. Note that the addition of new functions to the program
entails only the addition of new links to the skip chain and
linkage of the new subroutines, without further modifications of
the existing code.

The command interceptor and interpreter is programmed to
display the contents of two display arrays on the X-Y scope at all
times. These arrays can be filled with any of the functions that
need be displayed during operation (μ vs. E during data acquisition, the EXAFS spectrum vs. k or the Fourier transform vs. R
during analysis). For the program to be interactive, it should
spend most of the time within the command interceptor code. This,
of course, is the case during data analysis, since these commands
take but a few secnds to execute. In order to accomplish this
during data collection, which can take hours, the following
construction was adopted: Upon receipt of a scan command, the
instrument is readied, moved to the beginning of the scan range
and the scalers started. The program then jumps into the command
interceptor code and displays the data already collected (at this
point, no data are available yet). When the reference scaler
overflows, an interrupt is activated and the program automatically
enters an interrupt service routine, which reads the scalers,
computes and stores the measured data point, updates the display,
moves the spectrometer to the next point and returns. A flow
diagram of the interrupt service routine is shown in Fig. 2.

With this scheme, the program always displays the data
collected thus far, at the same time being able to accept and
process user input. An unsuccessul scan can be aborted very
conveniently in this manner, while still under program control.

At the present time, our software can perform a wide variety
of tasks. A number of utility operations are available to move
the spectrometer, change the set point of the sample cryostat,
etc. The data acquisition routines allow sufficient latitude in

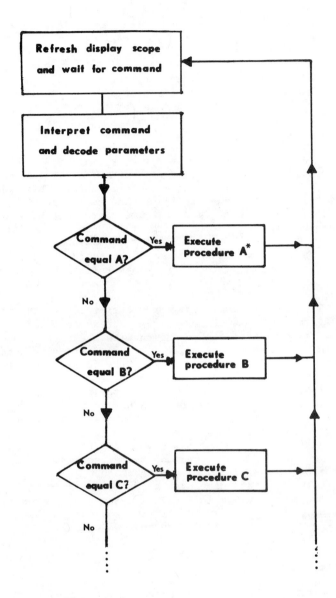

*Subroutine call unless procedure very simple.

Fig. 1. General layout of the main program.

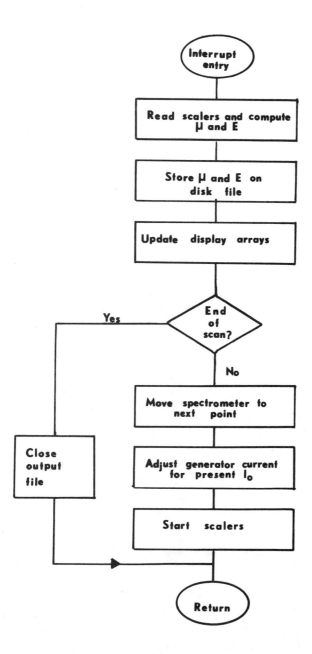

Fig. 2. Interrupt service routine for data acquisition.

the choice of scan parameters, generator control and data display. The analysis routines can perform background subtraction, computation of the EXAFS spectrum, Fourier transforms and filtering and data smoothing. A number of input/output routines can print, plot, save or retrieve any of the data arrays, automatically including the date and time and an optional user-chosen title. Finally, a simulation routine is available to compute EXAFS spectra of selected atomic models with up to nine atomic shells. Currently under development is a nonlinear least squares procedure, which will allow refinement of selected parameters of a model, with the capability of imposing constraints on the solution.

THE MONSANTO EXAFS FACILITY

The Monsanto "in house" EXAFS system will use a HP 9825S computer, interfaced to a HP 6942A Multiprogrammer for data acquisition and for control of the sample environment. Data reduction will share the facilities of a PDP 11/34A computer, operating under the RSX-11M Operating System.

The EXAFS data reduction and analysis software will be written in standard ANSI FORTRAN IV and will incorporate a Tektronix 4025 graphics terminal for fast inspection of the data. Hard copy output initially will be through a Versatec Printer/Plotter. Communication between the HP 9825S and the PDP-11/34A will take place via a RS232 interface.

The HP 9825S computer is programmed in HPL, an easy to learn, powerful interpreter language, similar to an extended version of BASIC. The computer has 23 kbytes of user random access memory and approximately 43 kbytes of read-only memory, containing the system monitor and the HPL interpreter. A special "live" keyboard feature simplifies data acquisition/control type program development by allowing any variable to be examined or modified while a program is running. Program modifications are very easy to implement and all statements are checked for syntax errors as they are entered.

The HP 6942A Multiprogrammer serves as an intelligent I/O device able to store and execute a series of commands that control the operation of up to sixteen user selected cards. The system has a real time clock and also memory, which may be used to buffer high-speed input/output operations.

The Multiprogrammer is interfaced to the controlling computer via a HP-IB (IEEE-488) interface bus. The monitors of some DEC-11 and HP-1000 computer systems have IEEE-488 drivers available, which can be called from FORTRAN programs. However, controlling the Multiprogrammer with FORTRAN code is not quite as convenient or as fast as with HPL running on the HP 9825S computer.

The Multiprogrammer unit in our system is currently set up with 16 bit digital input, output, interrupt, memory and programmable stepping motor controller circuit cards, as shown in Fig. 3. Additional custom logic will be added later to control the X-ray generator, service the counters and drive the display scope. This circuitry will be on breadboard cards which plug into the Multiprogrammer and use power from it.

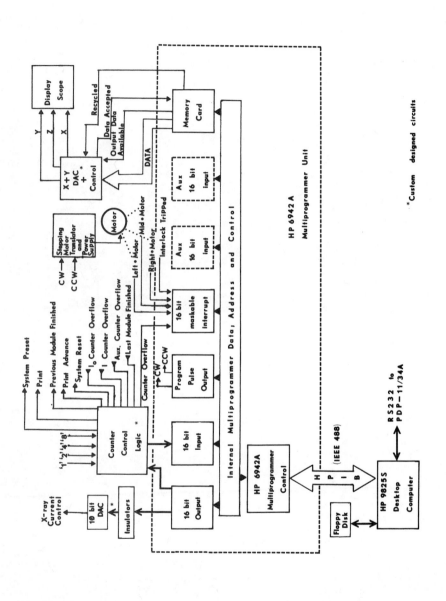

Fig. 3. Monsanto EXAFS data acquisition system.

EDITOR NOTES - CHAPTER 11

z. See also Chapter 5.

y. See also Chapter 7 and 12.

x. In Chapter 5 a description of the University of Washington Computer interfaced EXAFS system is also given.

RELATION OF ROLES OF IN-LABORATORY
AND SYNCHROTRON RADIATION EXAFS FACILITIES

Arthur Bienenstock
Stanford Synchrotron Radiation Laboratory, Stanford University
Stanford, California 94305

ABSTRACT

The paper begins with a technical assessment of in-laboratory systems utilizing conventional (tube) x-ray sources. Then a number of advantages of laboratory systems, compared to synchrotron radiation systems, are presented. These include rapid implementation of ideas, the setting of priorities by the experimenter, the initial training of students, the avoidance of travel and the performance of proprietary research. In contrast, synchrotron radiation offers significantly better resolution which may allow for the observation of complex coordination systems which are likely to be missed with the laboratory systems, the ability to study the coordination of quite dilute constituents and surface atoms, a broader photon energy range, access by those without the expensive laboratory systems as well as sophisticated education of students and their interactions with the outstanding x-ray scientists who use SR facilities. The paper concludes with some predictions and questions about the future relationships between the two types of facilities.

INTRODUCTION

In this paper, the most likely to develop relationships between in-laboratory and synchrotron radiation EXAFS facilities are discussed. The paper begins with my own technical assessment of laboratory systems, based on the papers presented at this meeting. Then it spells out my perception of the relative advantages of laboratory and synchrotron radiation systems and goes on to predict and raise some questions about the future.

TECHNICAL ASSESSMENT

It is my feeling, based on the talks presented here, that in-laboratory EXAFS systems utilizing conventional sources are best for transmission experiments in the range of 2.5 to 20 KeV. These experiments can be performed with an energy resolution which ranges from slightly below 5 to approximately 17 eV. To my knowledge, no significant improvement in resolution is anticipated at the present time.

Fluorescence experiments can also be performed with laboratory systems, but the dilution limits are, as yet, undetermined. As has been mentioned previously,[2] the polarization of synchrotron radiation offers a significant advantage here, since one can discriminate against scattered radiation by having the fluorescence detector view radiation coming parallel to the polarization vector of the incident beam. Since the radiation in a laboratory system is not polarized, there is still significant scattering in all directions. Given, as well, the lower intensity, I anticipate that the dilution limits in fluorescence experiments using laboratory sources will probably lag by 2 to 3 orders of magnitude in concentration behind those achievable with synchrotron radiation.

Finally, it seems extremely unlikely that surface EXAFS experiments will be performable with laboratory systems through either the partial yield or Auger techniques.

LABORATORY SYSTEM - ADVANTAGES

The most significant advantage that I see for laboratory systems is the rapid implementation of an idea. That is, one need not submit a proposal to a Proposal Review Panel and then wait for access to time at a synchrotron radiation facility. Instead, one can often perform the experiment immediately. This is extremely important when a person performing structure analysis is involved with people who are concerned with other aspects of the same sample. Too often, a person utilizing a synchrotron radiation facility provides a structural answer after the others involved with the material have gone on to other things.

In the same way, it is the experimenter, rather than the Proposal Review Panel, that sets the priorities for use of the instrument. This can be extremely important at an industrial lab where knowledge of an atomic arrangement may be of immense importance in pushing forward some technical development, but may not rank high on a scientific scale. Similarly, knowledge of such atomic arrangements may be important to an interdisciplinary program at a university, but still not meet the competition for time at a synchrotron radiation facility.

In-laboratory systems may also be much better for the initial training of students. Time at synchrotron radiation facilities is now so precious and travel from home so expensive that graduate students' advisors are pressed to assure that their students' experiments are almost perfect. There is little time for mistakes and the correction of mistakes which contribute so much to the development of a scientist. That is, while the experiment is underway a thesis advisor, post-doc or senior student may be obliged to work at a sophisticated level which is beyond the experience of the student who is to do the learning in order to assure that the experiment will work. In a laboratory system, one would allow the student to build up experience through the process of making and correcting mistakes.

Clearly, an in-laboratory system also offers a company the opportunity to perform proprietary research which it need not describe to anyone. There are no patent problems related to the performance of the work. This is a major advantage.

Finally, the in-laboratory systems negate the need for extensive travel to perform the experiments.

SYNCHROTRON RADIATION FACILITIES ADVANTAGES

On the technical side, there is no question that the high flux, smooth spectrum, high photon energies, collimation and polarization of synchrotron radiation offer significant technical advantages for the performance of EXAFS work. First of all, they can provide significantly better resolution. Such resolution offers the possibility of determining much more accurate coordination numbers since it allows for obtaining better information about second and more distant neighbors. Finally, the high resolution allows the accurate measurement and utilization of edge and near edge structure. This information is becoming extremely important as we understand better how to utilize it.

As I indicate previously, we may anticipate that synchrotron radiation facilities will be able to perform experiments on very much more dilute impurities and "dilute" metallic constituents of metalloproteins and other complex macromolecules in solution than can in-laboratory systems. Since many properties of materials depend on atoms which are present in very low concentrations, this is a significant advantage for synchrotron radiation facilities.

Along the same lines, surface EXAFS utilizing partial yield or direct Auger detection appears quite performable now at synchrotron radiation facilities. It is not all apparent that it will be performable with laboratory systems. The high polarization of the synchrotron radiation also offers the opportunity to make studies parallel and perpendicular to given directions, which is an advantage not only in surface EXAFS but in single crystal transmission studies.

These two opportunities afforded by synchrotron radiation, to determine atomic arrangements around dilute constituents and at the surfaces of materials which need not be crystalline, are among the most important and exciting features of EXAFS.

Finally, the large spectral range over which the synchrotron radiation has a smooth and high intensity spectrum offers access to many more elements than is possible with in-laboratory systems.

In addition to these technical advantages, there are other advantages which should be noted. First of all, synchrotron radiation facilities offer access to EXAFS to those without the experimental equipment. Since the cost of a laboratory system is well above $100,000 and often runs as high as $200,000, this feature is not insignificant.[y]

In addition, synchrotron radiation facilities offer state-of-the-art education to graduate students. Although the students must work under the trying conditions described previously, they have the opportunity to learn the most advanced and effective techniques, like fluorescence studies of very dilute constituents and surface-EXAFS, as well as edge and near edge studies.

Finally, synchrotron radiation facilities are places at which some of the best scientists in the world in the x-ray science field work. The mixing together of these scientists over long periods on the "floor" of SSRL has led to a remarkable flow of ideas. Graduate students get to meet and interact with many of these scientists.

Some evidence of this is the large number of papers which are co-authored by famous people in the field and graduate students who are not directly associated with these well-known people. Often, the former come from industrial laboratories or other universities.

THE FUTURE

As a result of these considerations, I anticipate that one or more commercial spectrometers for in-laboratory work will become available over the next few years. If this happens, we may anticipate that many experimental groups will develop in-laboratory systems for the performance of routine experiments. This will mean, in turn, that many graduate students, post-docs and senior scientists will get their primary EXAFS education in such laboratories, rather than at synchrotron radiation facilities. These groups will use synchrotron radiation for sophisticated EXAFS experiments and new structural technique development (e.g., anomalous scattering).

The people involved will go to synchrotron radiation facilities much better prepared to make effective use of them than they are now. That is, their in-laboratory experience will provide them with experimental capabilities and also better judgment as to what can and cannot be done with EXAFS.

I fear, however, that some who would go to synchrotron radiation facilities will fail to do so. That is, they will find it too convenient to perform "routine" experiments in the laboratory and will never do the high resolution, ultimate experiments that they really should perform to pin down the properties of their systems. If this turns out to be the case, then graduate students will fail to develop their own capabilities. In addition, the diversion of resources, both intellectual and financial, from synchrotron radiation may lead to less emphasis to the development of what I shall term the ultimate experiments. That is, we will have had a diversion of resources from "great" to "good". It is, I believe, the responsibility of thesis advisors to assure that this does not happen and of federal program officers to assure that grants to laboratories for the performance of in-laboratory experiments also have sufficient funding for travel to synchrotron radiation facilities.

SOME QUESTIONS

Before closing, I would like to spell out some of the questions on my mind concerning this relationship between the relative roles of in-laboratory and synchrotron radiation facilities. We should attempt to get some idea what effect the development of the National Synchrotron Radiation Light Source at Brookhaven National Laboratory will have on our ability to meet EXAFS needs through synchrotron radiation. That is, one can ask the question as to whether its establishment will negate the need for in-laboratory facilities. I don't think so.

Contrarily, one can ask whether an insufficient number of in-laboratory facilities will lead to pressure on synchrotron radiation facilities to do "routine" experiments rather than the "ultimate" experiments of which they are capable. I believe that is the case at the present time and hope that it will be alleviated by the establishment of NSLS, CHESS and in-laboratory systems.

Finally, one should ask how resources, both intellectual and financial, will be apportioned between synchrotron radiation and in-laboratory facilities. Somehow, a balance must be obtained so that there are sufficient in-laboratory systems to meet routine needs while the resources are also available for the development of the ultimate experiments at the synchrotron radiation facilities.

ACKNOWLEDGMENTS

The author is indebted to Keith Hodgson and Herman Winick for suggestions which led to a considerable improvement of this paper. This work was supported on part by the National Science Foundation (under Contract DMR77-27489 for the operation of SSRL).

APPLICATION OF EXAFS IN CHEMISTRY - PANEL PRESENTATION

John D. Baldeschwieler
Division of Chemistry and Chemical Engineering
California Institute of Technology, Pasadena, California 91125

I would like to illustrate how a new physical method such as EXAFS might be applied in a modern chemical research environment. Issues of substantial scientific and practical interest provide a strong motivation for chemical studies which typically include

- development of synthetic methods
- determination of molecular structure (including intermediates)
- study of chemical reaction mechanisms.

Developing new synthetic methods is rarely straightforward. It is usually necessary to experiment with new chemical reaction steps under a variety of conditions. Typically side reactions occur yielding unexpected products in complex mixtures with the desired products, as well as the starting meterials. Extensive work is required to identify the components of such mixtures, to purify the desired intermediates, and to refine the reaction conditions to improve the yield of the desired product.

A variety of modern physical methods are essential to making progress in this complex environment. Typical physical methods include nuclear magnetic resonance spectroscopy (NMR), mass spectroscopy (MS), IR, UV and visible spectroscopy, laser Raman spectroscopy, atomic absorption and emission spectroscopy, gas chromatography and gas chromatography in combination with mass spec (GCMS), liquid chromatography, and x-ray diffraction. Rarely does one physical method provide a definitive answer; typically fragmentary information from a variety of physical techniques in combination with traditional analytical and synthetic approaches leads to progress.

For each of the physical methods used in modern chemical research, a hierachy of instrumentation capability is usually available. In the NMR area, for example,

- NMR instrumentation of modest capability is directly available to the working synthetic chemist. He typically operates these instruments himself, and uses them, for example, to direct step-by-step synthetic work. NMR is used to provide clues as to the composition of crude mixtures of reaction products. A typical laboratory instrument of this class is the Varian T-60 which sells in the $30-50,000 range.
- Most large laboratories also have NMR instrumentation of greater capability. Such instruments are usually run by a specialist in the particular technique and are used for less routine measurements, such as NMR of nuclei other than protons as well as spin decoupling and temperature dependent studies. A typical instrument in this class is the Varian

XL-100 which sells in the $100-200,000 range.
- A small number of "state-of-the art" instruments are available typically in federally-supported regional centers. Such centers contain unique instrumentation such as high-field NMR spectrometers utilizing superconducting magnets. Typical instrumentation in the NMR field in such centers includes spectrometers operating at 300-600 MHz for protons, produced by Varian or Bruker. Such centers play a role which is analogous to that of SSRL.

A central feature to the successful use of a typical method such as NMR in chemical research is that the instrumentation is supported by a large and capable industrial base. Industrial firms are involved in the development and refinement of new methods, and in the maintenance and support of instrumentation in the field.

I expect that if EXAFS is to have a significant impact on chemical research, that its pattern of usage will have to have many of the features described above. The synchrotron radiation facilities, of course, provide instrumentation of the highest currently available capability. Conventional x-ray sources will always be at a disadvantage with respect to intensity compared with synchrotron sources. For very dilute samples and for testing new experimental concepts, the additional flux available at a national facility is essential. However, such instrumentation centers are likely to have an impact on only a few specialized research problems.

Even with the possible future availability of dedicated synchrotron radiation sources, it is unlikely that EXAFS can become an important addition to the instrumental methods available for routine chemical structural studies if experiments can only be carried out at a small number of national facilities. The additional information provided by an EXAFS study often might not justify the cost and inconvenience of travel, delay and uncertainty of writing a proposal for beamtime, and such routine applications would probably not compete well with new experiments which would demand the full capability of the dedicated source. Other complications which arise in working in national facilities, for example, on proprietary materials are well known.

A small group of users may actually be able to extract data more effectively with conventional x-ray sources than with a national facility. The quieter laboratory environment and automated facilities can make possible long periods of data collection. Long periods of time needed for careful setup, alignment, and execution of experiments are also often not well suited for the research environment at a national facility. This type of experiment is more easily accomplished in a laboratory in which the principal researchers have direct control over an experiment for extended periods of time.

Laboratory EXAFS systems will also be valuable as training devices since changes and errors can be made without inconvenience to a large number of other users. Students with extensive experience on the laboratory EXAFS equipment can be expected to derive maximum benefit from the limited time at national facilities. A laboratory capability is also useful to optimize conditions before time is committed on a national source.

If EXAFS is to become a significant addition to the physical methods available for structural studies in systems of chemical interest, it will have to be developed from a specialized technique available that only a few national centers to a hierachy of instrumentation including systems that can be used and supported in a typical laboratory setting. This will require the development of laboratory-scale instrumentation, with active participation of industrial firms, as well as continuing support of specialized EXAFS studies at national synchrotron radiation facilities. The mix of capability provided by both laboratory spectrometers and specialized instrumentation centers will be an essential feature to the success of the method

RELATION OF ROLES OF EXAFS FACILITIES IN LABORATORY AND AT SYNCHROTRON RADIATION SOURCES - PANEL PRESENTATION

B. Ray Stults
Monsanto Company, St. Louis, Missouri 63166

ABSTRACT

Synchrotron radiation, because of its unique properties of polarization, extreme high photon flux, and tunability, will continue to be the choice for EXAFS research. However, as has been discussed at this workshop, a great many EXAFS experiments can be performed successfully using alternate energy sources. This is particularly true where the primary focus of the research is to study chemical systems using EXAFS as an structural, analytical tool. With this in mind, we at Monsanto are developing an in-house EXAFS laboratory using as our x-ray source the Elliott GX-21 rotating anode x-ray generator. We plan to use our in-house facility for most of our EXAFS research but continue using the Stanford Synchrotron Radiation Laboratory (SSRL) for studies which, due to factors such as dilute concentration, can best be performed at SSRL. Below is a brief discussion of some of the factors which led to our decision to build an in-house EXAFS laboratory. While some of these factors may be unique to industrial research others are applicable to most researchers at the synchrotron radiation laboratories.

Proprietary research. This is the primary factor which influenced our development of the in-house facility. Current regulations at SSRL require disclosure of the results of studies on proprietary materials within one year after completion of the work. This, in many instances, is not acceptable. While this may change in the future at SSRL and other synchrotron radiation laboratories, it is certainly not clear how or when any changes in this policy will be instituted.

Proposal Review System. Many problems of interest to industrial researchers, while being very important to company, would not rate very high under the proposal review system and would therefore not receive beam time. Also, as frequently demonstrated throughout science, experiments which initially look rather routine, produce some of the most interesting results. These type experiments are also not likely to receive high proposal ratings.

Experimental Timeliness. The total time from experimental design, submission of the research proposal, and obtaining beam time is often 9-12 months. This is too long for many industrial research projects. With your own in-house facility, the experimenter is able to control the priority of research projects.

Chemist-Spectroscopist Interaction. In an industrial research environment the scientist doing the EXAFS research is often not the one doing the chemical research and compound preparation. There is usually interaction with several different research groups. An in-house facility will greatly facilitate the interaction between the various groups and will almost certainly result in better experimental results.

Use of the Remote Facility. Travel to a remote facility is expensive and often very difficult. This is particularly true when one must transport a great deal of experimental equipment to the synchrotron radiation laboratory. Another problem with use of the remote facilities is sample integrity. Many samples are air or moisture sensitive and spectral data must be obtained soon after preparation. Transporting samples half way across the country becomes very difficult in these instances.

ROLES OF LABORATORY VERSUS SYNCHROTRON RADIATION EXAFS FACILITIES - PANEL PRESENTATION

Edward A. Stern
Physics Department, University of Washington
Seattle, WA 98195

RESEARCH ENVIRONMENT AND EDUCATION

The time pressure at a national synchrotron radiation source such as SSRL is all pervasive. Because time on the source is so precious there is great pressure to measure as much as feasible because of the fact that one will not be able to use the source again for several months.

Under such an environment what is missing is the possibility of what I call interactive innovativeness, the interaction between the experimentalist and his apparatus leading to new ideas. A laboratory apparatus will introduce interactive innovativeness back into EXAFS measurements. I find it necessary, in doing my best physics, to have the opportunity to continually interact with the experiment. After making a set of measurements it is important to try to digest the results. These results may pose puzzles which the experimentalist can then think about and after coming up with tentative solutions should check by further experiments. In short, at synchrotron sources there is no time to think and the laboratory facilities will reintroduce thinking into EXAFS measurements.

As a professor, one of my main responsibilities is educating my graduate students. I have been quite dissatisfied by the level of research education that my students could gain at a synchrotron source. In fact, this dissatisfaction was the major motivation to build my laboratory EXAFS facility. At the synchrotron source my students were only "button pushers" and were not learning how to design experiments and build equipment. They had no opportunity to learn the experimental skills that can be obtained only by doing. They had little opportunity to learn by making mistakes on their own. The time pressures at synchrotron sources and the long hours which drug the brain are not conducive to thinking or trying out things at the spur of the moment. The measurements, by necessity, had to be done by teams of researchers in order to maintain a 24 hour per day presence and it was not always feasible for a student to measure his own sample.

Since a user at a synchrotron source has contact with the facilities only a small fraction of the time, he is not familiar with the inner workings of the EXAFS apparatus nor of the changes that may have recently been made. It is therefore possible that when the apparatus is malfunctioning a student, with his lack of experimental experience, would not be aware of the malfunction. What is worse from an educational viewpoint, he is not able to become familiar enough with the apparatus so as to gain the intuition of sensing when it is malfunctioning. He thus is missing the opportunity of maturing his experimental expertise and distinguishing between reliable measurements and nonsense. A laboratory facility will correct these educational defects.

In order for the national synchrotron sources to have the trained and experience personnel for optimizing their facilities, the present generation of graduate students must be adequately trained. Laboratory EXAFS facilities can provide such training.

ROLES

It is clear from this Workshop that a laboratory EXAFS apparatus using a fixed anode source is feasible covering the x-ray range from about 2.5 - 20 KeV with an energy resolution spanning 1 - 10 eV. The cost of such an eventual apparatus will be low enough so as to make such facilities widely distributed. The laboratory facility is as adequate as the synchrotron source for measuring EXAFS on samples that are concentrated enough to use the transmission mode within this energy range.

In order to be most cost effective in use of synchrotron radiation facilities, these facilities should concentrate their instrumentation resources on those EXAFS measurements that only they can do. The unique characteristics of synchrotron sources have already been listed in previous talks. These characteristics lead to the instrumental needs for fluorescent measurements on dilute samples, polarization studies, timing studies, studies requiring highly collimated x-rays and high energy resolution. In addition, synchrotron sources will be needed for filling all EXAFS needs outside the energy range of 2.5 - 20 KeV.[x]

The synchrotron radiation sources have the difficult responsibility of characterizing their facilities and maintaining them so that a casual user can measure what he thinks he is measuring.

EDITOR'S NOTES - CHAPTER 12

z. See discussion by Sandstrom and Fine in Chapter 10.

y. Using a fixed anode x-ray source, laboratory EXAFS facilities can be built for about $50,000. The costs quoted here are assuming a rotating anode source.

x. This statement refers to x-ray facilities. The electron microscope utilizing electron energy loss instrumentation as described in Chapter 6 can cover the energy range below 2.5 KeV, and in a laboratory setting, too.

APPENDIX - PARTICIPANTS IN WORKSHOP, APRIL 28-30, 1980

1. Mr. J. Azoulay, Physics Department, University of Washington, Seattle, WA 98195

2. Prof. John Baldeschwieler, Division of Chemistry and Chemical Engineering, California Institute of Technology, Pasadena, CA 91125

3. Dr. Dwight Berreman, Bell Laboratories 1d 435, Murray Hill, NJ 07974

4. Prof. Arthur Bienenstock, Stanford Synchrotron Radiation Laboratory, Stanford University, Stanford, CA 94305

5. Mr. John Boland, Division of Chemistry and Chemical Engineering, The Chemical Laboratories, California Institute of Technology, Pasadena, CA 91125

6. Mr. C. Bouldin, Physics Department, University of Washington, Seattle, WA 98195

7. Dr. B. Bunker, Physics Department, University of Washington, Seattle, WA 98195

8. Mr. G. Bunker, Physics Department, University of Washington, Seattle, WA 98195

9. Prof. Gary Christoph, Chemistry Department, 140 W. 18th Avenue, Ohio State University, Columbus, OH 43210

10. Dr. Gabrielle Cohen, National Bureau of Standards, Washington, DC 20234

11. Dr. Stephen Crane, Division of Chemistry and Chemical Engineering, The Chemical Laboratories, California Institute of Technology, Pasadena, CA 91125

12. Mr. S. Csillag, Physics Department, University of Washington, Seattle, WA 98195

13. Dr. W. Tim Elam, Physics Department, University of Washington, Seattle, WA 98195

14. Mr. Ronald Emrich, Department of Chemical Engineering, Center for Catalytic Science and Technology, University of Delaware, Newark, DE 19711

15. Mr. J. Feldhaus, Stanford Synchrotron Radiation Laboratory, Stanford, CA 94305

16. Dr. Ron Gamble, Division of Chemistry and Chemical Engineering, The Chemical Laboratories, California Institute of Technology, Pasadena, CA 91125

17. Dr. P. Georgopoulos, Argonne National Laboratory, B-212, 9400 S. Cass Avenue, Argonne, IL 60439

18. Dr. Bob Greegor, The Boeing Company, P. O. Box 3999, Seattle, WA 98124

19. Mr. William A. Grote, Monsanto Company, 800 N. Lindbergh, St. Louis, MO 63166

20. Prof. R. Haensel, Institut für Experimentalphysik der Universität Kiel, Olshausenstrase, 4o-6o, D-23oo Kiel 1, F. R. Germany

21. Dr. Gregory Hamill, GTE Labs, Inc., 40 Sylvan Road, Waltham, MA 02154

22. Dr. William Harris, Chemistry Division, National Science Foundation, 1800 G. Street, Washington, DC 20550

23. Dr. J. Hastings, National Synchrotron Light Source, Brookhaven National Laboratory, Upton, NY 11973

24. Mr. Guy H. Hayes, Physics Department, University of Connecticut, Box U46, Storrs, CT 06368

25. Dr. Steve Heald, Brookhaven National Laboratory, Bldg. 480, Upton, NY 11973

26. Mr. D. G. Hempstead, Rigaku/USA, Inc., 3 Electronics Avenue, Danvers, MA 01923

27. Mr. John Holbin, Marconi-Elliott Avionics, Ltd., Elstree Way, Borehamwood, England, U. K.

28. Dr. Elizabeth Holt, Department of Biochemistry, University of Georgia, Athens, GA 30605

29. Prof. R. Ingalls, Physics Department, University of Washington, Seattle, WA 98195

30. Prof. Dale Johnson, Center for Bioengineering, RF-52, University of Washington, Seattle, WA 98195

31. Dr. James R. Katzer, Center for Catalytic Science and Technology, Department of Chemical Engineering, University of Delaware, Newark, DE 19711

32. Mr. E. Keller, Physics Department, University of Washington, Seattle, WA 98195

33. Mr. K. Kim, Physics Department, University of Washington, Seattle, WA 98195

34. Dr. Gordon S. Knapp, Argonne National Laboratory, B-212, 9700 S. Cass Avenue, Argonne, IL 60439

35. Mr. K. Lu, Physics Department, University of Washington, Seattle, WA 98195

36. Dr. Philip J. Mallozzi, Battelle Columbus Laboratories, 505 King Avenue, Columbus, OH 43201

37. Dr. T. Matsushita, Stanford Synchrotron Radiation Laboratory, Stanford University, Stanford, CA 94305

38. Dr. Namavar, Physics Department, University of Connecticut, Box U46, Storrs, CT 06268

39. Dr. T. Oversluizen, National Synchrotron Light Source, Brookhaven National Laboratory, Upton, NY 11973

40. Dr. Donald R. Petersen, Dow Chemical Company, Bldg. 1776, Midland, MI 48640

41. Prof. Donald Sandstrom, Physics Department, Washington State University, Pullman, WA 99164

42. Prof. Dale Sayers, Physics Department, North Carolina State University, Raleigh, NC 27607

43. Dr. R. E. Schwerzel, Battelle Columbus Laboratories, 505 King Avenue, Columbus, OH 43201

44. Prof. E. A. Stern, Physics Department, University of Washington, Seattle, WA 98195

45. Dr. B. Ray Stults, Monsanto Company, 800 N. Lindbergh, St. Louis, MO 63166

46. Dr. Grayson Via, Exxon Research Laboratories, Linden NJ 07036

47. Prof. Y. Yacoby, Racah Institute of Physics, Hebrew University, Jerusalem, Israel

AIP Conference Proceedings

No. 45	New Results in High Energy Physics - 1978 (Vanderbilt Conference)	78-67196	0-88318-144-
No. 46	Topics in Nonlinear Dynamics (La Jolla Institute)	78-057870	0-88318-145-
No. 47	Clustering Aspects of Nuclear Structure and Nuclear Reactions (Winnepeg, 1978)	78-64942	0-88318-146-
No. 48	Current Trends in the Theory of Fields (Tallahassee, 1978)	78-72948	0-88318-147-
No. 49	Cosmic Rays and Particle Physics - 1978 (Bartol Conference)	79-50489	0-88318-148-
No. 50	Laser-Solid Interactions and Laser Processing - 1978 (Boston)	79-51564	0-88318-149-
No. 51	High Energy Physics with Polarized Beams and Polarized Targets (Argonne, 1978)	79-64565	0-88318-150-
No. 52	Long-Distance Neutrino Detection - 1978 (C.L. Cowan Memorial Symposium)	79-52078	0-88318-151-
No. 53	Modulated Structures - 1979 (Kailua Kona, Hawaii)	79-53846	0-88318-152-
No. 54	Meson-Nuclear Physics - 1979 (Houston)	79-53978	0-88318-153-
No. 55	Quantum Chromodynamics (La Jolla, 1978)	79-54969	0-88318-154-
No. 56	Particle Acceleration Mechanisms in Astrophysics (La Jolla, 1979)	79-55844	0-88318-155-
No. 57	Nonlinear Dynamics and the Beam-Beam Interaction (Brookhaven, 1979)	79-57341	0-88318-156-
No. 58	Inhomogeneous Superconductors - 1979 (Berkeley Springs, W.V.)	79-57620	0-88318-157-
No. 59	Particles and Fields - 1979 (APS/DPF Montreal)	80-66631	0-88318-158
No. 60	History of the ZGS (Argonne, 1979)	80-67694	0-88318-159-
No. 61	Aspects of the Kinetics and Dynamics of Surface Reactions (La Jolla Institute, 1979)	80-68004	0-88318-160-
No. 62	High Energy e^+e^- Interactions (Vanderbilt, 1980)	80-53377	0-88318-161
No. 63	Supernovae Spectra (La Jolla, 1980)	80-70019	0-88318-16.
No. 64	Laboratory EXAFS Facilities - 1980 (Univ. of Washington)	80-70579	0-88318-16
No. 65	Optics in Four Dimensions - 1980 (ICO, Ensenada)	80-70771	0-88318-16